Gathering Electr
Waste in Tanza

Gathering Electronic Waste in Tanzania

Labor, Value, and Toxicity

Samwel Moses Ntapanta

LEXINGTON BOOKS

Lanham • Boulder • New York • London

Published by Lexington Books
An imprint of The Rowman & Littlefield Publishing Group, Inc.
4501 Forbes Boulevard, Suite 200, Lanham, Maryland 20706
www.rowman.com

86-90 Paul Street, London EC2A 4NE

British Library Cataloguing in Publication Information Available

Library of Congress Cataloging-in-Publication Data

Names: Ntapanta, Samwel Moses, author.
Title: Gathering electronic waste in Tanzania : labor, value, and toxicity / Samwel Moses Ntapanta.
Description: Lanham, Maryland : Lexington Books, 2025. | Includes bibliographical references and index. | Summary: "Using Tanzania as a case study, this book traces the lifecycle of electronic goods beyond their discard in the Global South to provide a deeper understanding of the informal sector's vital role in rethinking global discard circulations, labor, value, and toxicity"–Provided by publisher.
Identifiers: LCCN 2024044415 (print) | LCCN 2024044416 (ebook) | ISBN 9781666956269 (cloth) | ISBN 9781666956276 (epub)
Subjects: LCSH: Electronic waste–Tanzania—Dar es Salaam. | Scrap metal industry–Employees–Tanzania—Dar es Salaam. | Scrap metal industry–Environmental aspects–Tanzania–Dar es Salaam. | Informal sector (Economics)–Tanzania–Dar es Salaam. | Dar es Salaam (Tanzania)–Social conditions.
Classification: LCC HD9975.T343 D3 2025 (print) | LCC HD9975.T343 (ebook) | DDC 363.728809678–dc23
LC record available at https://lccn.loc.gov/2024044415
LC ebook record available at https://lccn.loc.gov/2024044416

"Keep telling the other story, the untold one, the life story. It matters what story we tell to tell other stories with; it matters what concepts we think to think other concepts with. It matters where how ouroboros swallows its tale, again. That's how worlding gets on with itself in dragon time."

Donna Haraway
(Le Guin, Haraway and Lee 2019, 10).

Contents

Acknowledgments

First and foremost, writing this book wouldn't have been possible without my *walimu* (teachers), who allowed me into their lives and let me share their life stories with the world. I am forever indebted to everyone at Mahakama Ya Friji, Uhuru Peak Glacier, and Uhuru Street in Kariakoo.

The study that led to this book was funded by the University of Oslo: Life Science funds for convergence environments—interdisciplinary research groups that address major challenges within health and environment. A special debt of gratitude is owed to my colleagues at the Anthropology of Toxicity (AnthroTox) project, with whom I have collaborated throughout my study: Prof. Paul Wenzel Geissler, Prof. Ruth Jane Prince, Prof. Theodoros Rakopoulos, Dr. Ane Haarr, Christian Medaas, Signe Mikkelsen, Dr. Franziska Klaas, and Dr. Maja Nipen. I also extend warm thanks to all members of the AnthroTox project: Prof. Katrine Borgå (University of Oslo, Biosciences), Prof. Knut Breivik (University of Oslo, Chemistry), Prof. Susanne Bauer (University of Oslo, Technology, Innovation and Culture), Prof. Aviti John Mmochi (University of Dar es Salaam), Dr. Peter Mangesho (Tanzania National Institute for Medical Research), Dr. Vendelin Simon (University of Dar es Salaam), Prof. Anders Ruus (University of Oslo, Department of Biosciences), Prof. Martin Schlabach (Norwegian Institute for Air Research), Prof. Jan Ludvig Lyche, and Prof. Anuschka Polder (Norwegian University of Life Sciences (NMBU)).

I also extend thanks to the Tanzania Commission for Science and Technology (COSTECH) and the Tanzania National Institute for Medical Research (NIMR) for ethical clearance of this study. I would also like to extend my gratitude to the University of Dar es Salaam, where my journey started as an undergraduate student and for hosting me during my fieldwork.

Extending thanks notes to the students, administration, and academic staff at the University of Bayreuth. During my two-year post-doctoral placement, I wrote the manuscript for this book. Warm thanks to the Department of Social and Cultural Anthropology, the Chair of Social and Cultural Anthropology with a focus on Africa under Prof. Andrea Behrends, the African Multiple Cluster of Excellency, and the Institute for African Studies. Special gratitude to Ms. Janine Nagat, Prof. Jia Hui Lee- Science & Technology Studies of the Global South, Prof. Benjamin Kirby—Study of Religion with a focus on global entanglements, Prof. Joël Glassman—History with a focus on African history, and Dr. Jochen Lingelbach. Thanks to Prof. Steffen Köhn and Dr. Jonalou Labor, members of the social life of cryptocurrencies: How Digital Currencies Impact Informal Economies in the Global South project and the whole Department of Anthropology at Aarhus University.

I dedicate this paragraph to exceptional people who have been hand-in-hand with me throughout the study and writing of this book. Dr. Pierre du Plessis and Prof. Heather Swanson, thank you very much for always being there. To all my friends who have helped me so much during this journey, a few to mention are Dr. Edwin Emeso, Salim Mwanakatwe, Deogratius Wilbad Mtei, Amani Mtei, Godfrey Mtei, Francis Chukwu Chiduo, Grace Nanyaro, and Aloyce Ndyamukama.

Very special thanks go to Moe, Evy, and Martha. I cannot describe how grateful and happy I am to have you. And to the whole Mouritsen family, for the support and care. To my parents, Annastazia Sungura and Moses Samwel Ntapanta: there are no words to thank you. My siblings, Catherine, Charles, Claudia, and Yohana, you are always in my heart. Finally, to my grandmother, Mdala Bilha Balisidya. For all the love you have shown me, thank you. And to my namesake, Samwel Ntapanta, continue resting in power.

All observations were done between March 2018 and October 2023. The author's reflections and narrations are based on historical sources and personal memoirs. Because of this, there is back-and-forth use of the present and past tense. The author is responsible for all translations from Swahili to English. Swahili words appear in italics on first use, followed by their translations in parentheses. Further information is placed in the endnotes where clarification about translated words is needed. Some parts of the book were published in *Anthropology Today* journal, published by John Wiley & Sons on behalf of the Royal Anthropological Institute and Norsk Antropologisk Tidsskrift. Currency exchanges between the Tanzanian shilling (TZS) and U.S. dollar (USD) are converted on the average exchange rate in 2019—1 USD for 2,307 TZS.

Introduction

The Workshop at the Downstream of Techno-Capitalism

"WHERE IS E-WASTE?"

In March 2018, I went on a pre-fieldwork trip to Tanzania with my colleagues from the Anthropology of Toxicity (AnthroTox) project at the University of Oslo. The team's imagination on this journey was fueled in equal measure by the statistics about enormous amounts of e-waste being exported from Europe to Africa and by the powerful images of Western media reports on illegal waste dumping on the continent, and in particular, unregulated electronics recycling. During this trip, we expected to find landscapes resembling those depicted in the 2018 documentary *Welcome to Sodom*, directed by Austrian filmmakers Florian Weigensamer and Christian Kröhnes (Weigensamer and Krönes 2018). The film shows Ghanaian e-waste workers at Agbogbloshie, a world-famous e-waste recycling center. Many are in their youth, interacting with fire and smoke to extract copper and other metals from e-waste. The nearby river is heavily polluted by the activities. At the documentary's opening, a young person raps about the place being the Sodom, seemingly of e-waste.

Although on the trip we did observe enormous amounts of new and (often well-) used electronics in the popular Kariakoo, a name derived from the mispronounced post-World War II term "career corps," and Darajani markets in Dar es Salaam and Unguja, respectively, we were surprised by the amount of e-waste arriving in the cities' landfills. We did observe plastic used in electronics: polystyrene, polypropylene, or acrylonitrile butadiene styrene at the landfills. However, there were few signs of unregulated e-waste activities around landfills where we expected to find them. The fire, the burning, and blackbodies laboring that represent unregulated e-waste recycling in Africa were not captured by our searching gaze. Where was the e-waste that we had come to study?

There are explanations for why the team, particularly me, found no significant e-waste activities at the landfills. The idea I had from readings about "informal" e-waste recycling in Ghana would resemble, to some extent, to the activities in Dar es Salaam. Even though I lived in Dar es Salaam for six years between 2006 and 2012, I had not visited the landfill or followed up on e-waste or waste management infrastructures in the city. The landfill visit was influenced by the thought that e-waste activities would be around the landfill. There was also little literature and few reports about the e-waste situation in Tanzania. An e-waste assessment for Tanzania commissioned by the UNIDO e-waste initiative in 2011 was the only report available. The report described the e-waste ecology in Tanzania, which is dominated by the informal recycling sector (Magashi 2011). However, the report lacked substantial information on where and how e-waste activities occur. Of course, the landfills became the first place to start.

There was also a conceptual setback. The conceptualization of discarded electronics as e-waste led me down a rabbit hole. This thinking describes well why I first went to the landfill. The term e-waste carries a certain inference. Concepts matter and influence how we approach, theorize, and produce knowledge about humans and non-humans. Concepts familiarize the strange (Liboiron and Lepawsky 2022). The concept of waste creates certain underlying social, cultural, and economic assumptions. My search for e-waste in Dar es Salaam was entangled with the assumption of waste as a matter out of place and undesirable. Because concepts matter, throughout this book, I will use the two concepts interchangeably, e-waste to mean that which is out of place like toxic compounds released during recycling and electronic discards to mean that which is discarded does not "inexorably become waste" (Gregson 2023, 31). The concept of electronic discards (e-discards) opens up a broader conceptualization of what e-waste is and what acts of wasting electronics might include. It also allows us to defamiliarize e-waste (Liboiron and Lepawsky 2022).

Unregulated electronic discards labor in the Global South, often called informal e-waste recycling, has been portrayed as catastrophic, a re-creation of biblical apocalypses. I will use, in some instances, the term informal to mean unregulated. As the epigram from Donna Haraway at the beginning of this book warns about the story we want to tell and what concept we use, I narrowly adopt the concept "informal" to mean activities that are not directly regulated by the state through taxation, non-salaried employment, flexible qualifications to enter and leave the labor market, recruitment based on trust, ethnic affiliation, and belonging.

My subsequent ethnographic quest to locate where electronic discards end up led me to the *Mahakama Ya Friji*, which translates as a refrigerator courthouse or workshop. The name Mahakama Ya Friji reflects the activities at this place. Electronic discards and scrap materials collected around the city are brought here to be salvaged. Metals like copper, cast iron, iron, and batteries are extracted to feed manufacturing demands. Other materials, like aluminum, are transformed on-site into various objects like charcoal stoves and roofing materials. These intricate and innovative recycling, upcycling, and reuse methods explain why the AnthroTox team found so few electronic discards in landfills.

Places like Mahakama Ya Friji are vital to Dar es Salaam's urban metabolism and livelihoods. At the same time, the workshop is connected to technological changes, global streams of discards, and value cycles. Economies and daily lives are increasingly connected to discards and waste: production, sorting, reusing, and recycling, or removing discards from the view (O'Neill 2019). Just as industrial workplaces generate global discards flows, waste streams become workplaces (Doherty 2021; Doherty and Brown 2019), places for value recovery and the reintegration of waste materials into new value cycles. This book focuses on understanding labor, value, and other relations assembled at downstream of techno-capitalism that connect to the materialities of electronics and e-discards, toxicity, and the city's history, social-cultural and economic fabrics.

Communities in and around discards workplaces like Mahakama Ya Friji live downstream of techno-capitalism. They are often the recipients of discards from the metropoles. Here, the metropole-periphery division not only connotes developed vs. developing countries, but also points to the cores and peripheries within countries, or within cities. Upper- and lower-class areas form their own core-periphery relationships within cities. In this case, electronic discards provide a crucial lens to explore waste and wasting, valuation and devaluation, and broad systems and power dynamics embedded within the politics and economies of discards (Liboiron and Lepawsky 2022).

Discards worksites, like the Mahakama Ya Friji, in the Global South are perceived as zones of disposable people and disposable materials, which contrasts with the seemingly clean, green, and proper ways and technologies of disposing of discards. The same contrast applies between affluent and poor neighborhoods within the Global South(s) or between elite and non-elite classes in cities. Even in the very same low-income areas, this classification continues.

To explore why the AnthroTox team could not find electronic waste at the city's landfill is to observe the city, the workshop, the repairer, the tools used, the objects dismantled and produced, the transactions, and the whole value-gathering process. These gatherings of relations connect to each other and to more extensive networks beyond the workshop, the city, and the country. Throughout this book, I describe the labor and knowledge of workers at resource frontiers, who expose themselves to multiple health dangers to recycle value embedded in electronic discards. The activities of workers at these sites profoundly reveal their vital role in extracting raw materials, making a livelihood, and their position in microeconomies and sociocultural assemblages. Although mainstream waste management finds their methods and mechanisms for recycling electronic discards unpalatable, this book pays close attention and seriously considers workers' vital position in addressing the electronic discards problem.

HAPPY TO DO IT

In October 2018, I was introduced to Mahakama Ya Friji. At the workshop, I met my *walimu*[1] (teachers). There are three groups of workers at the workshop: scrap gatherers, crafters, and scrap dealers. I remember my first day at the workshop. From a distance, a building fully covered with corrugated iron sheeting was visible, with piles of scrap materials scattered around it, and from within emanated banging sounds, loud voices, and smoke. Standing a few meters from the entrance to wait for my host George, a charcoal stove crafter and one of the workshop leaders, I could see men seated on old car seats with pieces of rails positioned between their feet, hammers in hand. As they hammered away, I could attribute the sounds I had heard meters away while approaching the workshop.

The sheer effort by one crafter in his 50s, nicknamed "Uncle Mjomba" (Mjomba Mjomba), bearing a blue T-shirt with the inscription *Happy to Do It*," became a fixation of mine. The satirical expression inscribed on his shirt signals global consumption patterns and their connections to his labor. As he effortfully banged the aluminum sheet to recast its shape, sweat streamed off his forehead. As he picked up his near-perfect rhythmic hammer movements, a sense of endurance and years of laboring could be deduced. I witnessed the precision of the cuts, edges, and shapes he produced. His mastery and competence in his work are almost unquestionable, as he delivered a charcoal stove within an hour of my conversation with George.

In Dar es Salaam, there are hundreds of small e-scrap backyard workshops like the Mahakama Ya Friji. Although these workshops are scattered all over the city, they share certain features: their locations, innovations and ingenuity, labor arrangements, and hybrid technologies. The labor processes in these places are flexible. Kinship and ethnic relations are sources of labor assemblages, ethical and moral guidance. These workshops are usually located in small, open backyard spaces in dense neighborhoods, typically sandwiched between other small businesses. Their locations are negotiations between visibility to customers and minimal interaction with state apparatuses. The workshops form synergistic relations with other small businesses that become part and parcel of discards economies. Food vendors, kiosk owners, repairers, and beverage sellers tend to surround workshops. The workshops become vibrant economic spots that connect to local, national, and global economies and discards streams. These are workspaces where people innovatively use local knowledge, skills, tools, and networks together with new technologies to transform discards objects into value (Mavhunga 2014, 2017).

Such workshops are difficult to spot. Unlike the former well-known electronic waste districts of Agbogbloshie in Ghana or Guiyu in China, in Dar es Salaam, they occupy pockets of the cityscapes, allowing visibility only to those working or doing business with them. A study conducted in 2018 in the Kinondoni municipality, one of five municipalities that form the Dar es Salaam metropolitan area, counted more than 85 workshops with more than 2,125 people working in them (Kyessi and Omar 2018).

I spent around three months before arriving at Mahakama Ya Friji. The workshop is located in Mwananyamala, a bustling residential area in Kinondoni in the middle of the city famously known as "*Uswahilini*," which figuratively means where the lower class lives. Five kilometers west toward the Indian Ocean coast lies the city center, occupied by offices of the government, embassies, financial institutions, and nongovernmental organizations. The Msimbazi River Valley separates the city center and Kinondoni. On one side of the valley is Kinondoni's low-income housing; on the other side stand gigantic skyscrapers penetrating the city center sky. To the northeast of Kinondoni is Mikocheni, Mbezi Beach, and the Msasani Peninsula, which host Msasani, Masaki, and Oyster Bay, the wealthiest neighborhoods in Tanzania. This interlocking proximity between Kinondoni and the Msasani Peninsula, respectively, forms the periphery and metropole relations within the city.

Workshops like Mahakama Ya Friji are gatherings of histories and labor, ethnicity and kinship relations, valorization and value transformation, and ingenuity and innovation. At the workshops, "traditional" techniques and late-capitalist technologies meet to address waste problems, create livelihoods, and foster entrepreneurship. These networks make discards work sites a collection of complex gatherings. Though these spaces, in many cases, are placed out of sight as the "other," they form networks of waste metabolism and gather vibrant social-political, economic, and biological lives.

By staying with Mahakama Ya Friji, the book focuses on labor and life downstream of techno-capitalism. Living downstream means living at the bottom of the capitalist system, in places that are at the receiving end of the violence of capitalism, where surplus and disposable populations live. To reveal the life and the downstream, the book follows labor in unregulated e-discards salvaging and lifescaping initiatives in response to exposures to

Figure 0.1 A Remade Photo of the First Day Encounter of Mjomba at the Workshop. Photo provided by the author.

hazardous chemicals. Lifescaping describes everyday practices for managing toxic exposure in the absence of proper information, tools, safety procedures, or regulation (Edelstein 2003; Tironi 2018). Lifescaping shows that even when communities are at the downstream, they have agency in their own lives, and they take the measures necessary for life to continue. All these practices appear in every aspect of life downstream of capitalism. Learning from unregulated and peripheral spaces offers new lessons for addressing major global issues like electronic discards.

THE DOWNSTREAM OF TECHNO-CAPITALISM

Electronic discards pose a significant challenge in African cities, particularly in Dar es Salaam where my field research was conducted. The city's inadequate waste collection and management infrastructure has given rise to unregulated workshops like Mahakama Ya Friji. These workshops use the infrastructure gap to handle e-discards, operating within regulatory frameworks that do not explicitly address the toxicity, thereby potentially endangering both the environment and public health.

The well-known story of unregulated value extraction activities from e-discards is that they perpetuate the spread of hazardous chemical compounds in the bodies of recyclers and the environment. These compounds are released during salvaging, thus exposing workers, surrounding dwellers, and the environment. The coming together of toxicants and violence happens here at the downstream end of capitalism. Primitive accumulation has not stopped, and the savage side of capitalism is experienced by those living "downstream" (Luxemburg 2003; Steingraber 2010; Tsing 2015a). When toxic compounds are released, they gather and come together in the environment and in bodies. Chemical compounds like persistent organic pollutants (POPs) from e-discards have different lifespans, some shorter while some persist for longer periods. The exposure to these compounds produces wastelands and wasted bodies.

The debate over pollution and toxicity was for a long period mainly conducted within the corridors of natural sciences like chemistry, biology, and technology on one side, and politics (which actors do what, when, and how in terms of producing, exacerbating, and concealing, as well as regulating, minimizing, and abating risks) on the other. The entrance of the social sciences, in this case anthropology, has brought the human and recently the nonhuman into the foreground, focusing on experiences of toxic harm, on everyday lives in polluted worlds, and on forms of knowledge, action, and activism. Beyond their chemical compositions or the harm they can cause, toxic compounds also have social and cultural significance (Beck 1992).

Recent convergences between the natural and social sciences, like the AnthroTox project through which this book was born, have bridged environmentalism, environmental justice, and the environmental humanities. They have revealed impending and ongoing environmental disasters that not only affect vulnerable communities, but also the planet as a whole, and have opened up avenues to push for change (Brown 2007). Communities and activists, with the backup of scientific evidence, can push politics and capital to do what they are supposed to do when it comes to containing and regulating toxic compounds, compensating affected communities, and cleaning up polluted environments (Brown 2007). However, activism and scholarship have underrepresented what people actually *do* when they become aware of their exposure but have little or no power to evoke changes to the situation, or when changes take a long time to occur (Tironi 2018; Tironi and Rodríguez-Giralt 2017; Liboiron, Tironi, and Calvillo 2018).

However, if we pay attention and learn from unregulated activities that address the e-waste problem without waiting for high-tech interventions, we may find the solutions we are seeking. This book comprises an ethnographic contribution to observing, learning, and understanding why things happen, how they happen, and records the effects of unregulated mechanisms, which can help address the e-waste problem. Practices around defunct electronics (i.e., refurbishing, recycling, reusing, and repairing) envisage a future through ingenuity and creativity that gradually creates grounds for entrepreneurship and livelihood creation. These "micro activities" are valuable in building the city and the world.

Dar es Salaam is one of the fastest-growing metropoles in the world. Nevertheless, it is located at the receiving end of value, at the downstream of techno-capitalism. Large quantities of metropoles' devalued objects can be found in markets and shops around Dar es Salaam. Secondhand goods are everywhere, from shoes, clothes, bags, bicycles, TV, mobile phones, refrigerators, and cars to big machinery and even hospital equipment (Malefakis 2019). The prevalence of Japanese cars, secondhand clothes, and computers from Europe and the Americas is evidence of how big the secondhand market is in Tanzania. Of course, there are reasons why the secondhand market is predominant, as I will describe in detail in chapter 2.

Nevertheless, the widespread demand for secondhand goods and the desire to have those goods are not only caused by poverty or lack of income. Of course, these two factors contribute to that, as I will show in the coming chapters. However, there is more than just poverty. The quality of new goods, in this case, electronics, arriving from China and other newly emerging economies does not reflect what people aspire to, compared with the goods produced for Western markets. The poor quality results from poor regulatory enforcement on the Tanzanian side. On top of that, due to many years of colonial and imperial manufacturing relations, Africa, since the colonization in the late nineteenth century to this day, has been the destination for used products from the metropoles for reselling or in the form of aid and donations. These relations have created a belief, with poor quality products from emerging manufacturers, that imported used electronics from the West are of better quality.

These used and, in many cases, worn-out devices will make several trips to repairers where their lifespan will be extended for some time. Nevertheless, they will reach the end of their life sooner or later. The Dar es Salaam city council, along with its consulting companies, collects less than 40% of the total waste produced in the city (Kaseva and Mbuligwe 2005). Many of these objects, now discards, will end up in dumpsites, both formal and informal, in river valleys, and some because of their materiality, like electronics, will be picked up by collectors.

The downstream is also a sacrifice zone of techno-capitalist accumulation (Little 2023, 2017). As a result, Tanzania is not just the end destination for devalued devices; simultaneously, primitive accumulation is also happening. The accumulation of capital from e-discards recycling in Tanzania emerges in two folds: On the one hand, it is a re-enactment of the enclosure of the agrarian system, pushing peasants from the countryside into cities to provide the labor needed by capitalism. This has its history and first happened during colonialism, through taxation and the introduction of colonial money, and later under liberalization, Structural Adjustment Programs (SAP), the introduction of the market economy, and the concentration of essential social and economic infrastructures in urban centers (Li 2010; Nguyen 2019). Tanzania's peasant economy collapsed from the early 1990s because of exploitative crop pricing in global markets and subsidy removals under austerity measures. In the meantime, peasants face land grabbing for commercial crops, some for the production of green energies including biofuel, solar, and wind plants (Greco 2015; Nelson, Sulle, and Lekaita 2012; Purdon 2013; Sulle and Nelson 2009). In this way, spheres of production and spheres of consumption that hold the peasant economy are disembedded from each other (Polanyi 1944). As a result, people have no means other than moving to the cities for wages or joining the informal labor market.

On the other hand, it is the continuation of colonial relations whereby Africa, even in the twenty-first century, remains a site of primitive accumulation, both in the appropriation of raw materials, cheap labor, and the market for cheap and discarded goods. The disposable peasant labor, driven into cities, is exploited for the purpose of capital accumulation, and specifically deployed for the extraction of raw materials for the global market. In cities, expelled peasants find no wage labor, access to social security, or health insurance. They ingeniously have to work under extreme conditions to earn a living and provide labor for the capitalist system. Recycling, and in this case e-discards, opens a whole new platform for the continuation of exploitation, plundering, and economic subordination of the Global South(s) by the metropole(s).

These measures: peasants' enclosure, discarding electronics, and continued colonial relations, are feeding the capitalist spirit by expanding to the peripheries not only to quench its thirst for profit or outdo the competition but because of forces within global capital and the declining of opportunities in the metropole. Underconsumption, economic stagnation, and the increased cost of production, taxes, labor, and environmental regulations on the one hand, and, on the other hand, human rights, environmental justice, and activism in the metropole force capital to the peripheries (Luxemburg 2003).

However, it does not mean that peasants who migrate to cities are fully integrated within the capital system. They don't have job contracts when migrating. Most of them will not be able to find wage labor. However, they come with relations outside the capitalist system like ethnicity, lineage, and race that are vital for capitalist accumulation from the margins. The notion of "peri capitalist" spaces (Tsing 2015a) aptly describes the nature of places like the Mahakama Ya Friji workshop. The workshop is a peri capitalist space. A space outside and at the same time inside the capitalist system, where social relations outside the capitalist system like ethnic and kinship relations, traditional knowledge, and people's ingenuity produce wealth for capital.

Most unregulated recycling in the African context organized through kinship and family relations, adapting traditional techniques and "rudimentary" tools, rely on traditional methods, technologies, and modes of organization such as kinship and ethnic relations that the capitalist system camouflages for accumulation. These relations become avenues for salvage accumulation (Tsing 2015a), in which capitalism concentrates wealth from conditions and processes outside the capitalist system, "taking advantage of value produced without capitalist control" (Tsing 2015b, 63). In contrast to primitive accumulation, salvage accumulation refers to a distinctly late-capitalist, twenty-first-century phenomenon whereby wealth accumulation occurs by turning noncapitalist relations into possibilities for wealth creation. Accumulation through salvage practices instantiates new forms of exploitation and creates different material relations. The capital is masked in social and cultural relations for further accumulation. However, salvage accumulation bears no difference to primitive accumulation when it comes to exploitation and violence. Salvage mostly turns savage, "translating violence and pollution into profit" (Tsing 2015b, 64).

The most violent feature of capitalism has been the creation of wastelands, where the life of humans, ecology, and nonhumans has to be wasted to produce (Doherty 2021). Those who live and work in wastelands are at

Figure 0.2 Mahakama Ya Friji Workshop. Photo provided by the author.

the downstream end of capitalism. They are trapped between their livelihood needs and the brutality of late capitalism, becoming recipients of a diminishing value chain, discards, and toxicants, that is, the ruins and rubble of capitalism (Gordillo 2014; Steingraber 2010; Stoler 2008). Capitalism was built and depends upon wasting and ruination (Stoler 2008). Tales of the progress of empire encompass continual degeneration, violence, and exploitation, all of which are deeply embedded in capitalist accumulation tendencies at the downstream end (Moore 2016; Tsing 2015a). However, the same system that destroys, produces, and then remakes its rubble again to continue to reproduce itself (Gordillo 2014). E-discards recycling is remaking the rubble or the ruins of capitalism into wealth accumulation. Unregulated e-discards recycling in Tanzania is another form of salvage accumulation that reveals how capitalism never ceases to concentrate wealth from any condition.

SEEING THROUGH A GATHERING LENS

This book opens with Donna Haraway's insightful words in Ursula Le Guin's book, *The Carrier Bag Theory of Fiction*. Haraway and Le Guin insist on telling the other stories, the untold and not the ubiquitous story. Man the Hunter was a story of patriarchy that Haraway and Le Guin are resisting only to tell. Inspired by this call, I have devoted myself, throughout this book, to telling the multiplicity of stories of unregulated e-discards recycling in Dar es Salaam, Tanzania. As Le Guin put it forward, there were many stories to tell than the story of an arrow penetrating the eye or the heart of the beast. The story was choreographically made popular as the protagonist story, and it still is. The other stories, the not-famous ones, are the stories I want to tell. The multiple ways of innovative practices of discards workers and their collectivities often remain hidden from view in the dominant narratives about electronic discards. In this book, I describe the innovative practices of discards workers in Dar es Salaam, not as a normative story that focuses on depicting them as perpetrators or victims of pollution. Their practices are rendered improper and marginalized by the state and even the environmentalism community, yet they are critical to local livelihoods and their social networks. The stories presented here aim to bring forward practices and the agency of discards workers as important actors against the backdrop of those normatively celebrated in late capitalism, which are driven by liberal logics of individualism and material accumulation.

Narratives about e-discards recycling are worlding stories. High-tech industrial recycling, concentrated in the Global North and monopolized by capitalists' conglomerates, is the proper, cleaner way to recycle electronic discards. Recycling from the margins, in places like Mahakama Ya Friji and

Agbogbloshie, is dangerous to health and the environment. There are two ways to look at these narratives. First, they are mainstreamed to create a story of heroes of e-discards recycling. Recycling e-discards is for high-tech companies; informal recyclers are polluters. Recycling conglomerates push for specific ideas, like the dangers of unregulated recycling. These narratives are purposefully stimulated to exclude certain recycling practices as dangerous and the other as pure, good, and sometimes holy (Douglas 2003). Second, corporations try to monopolize the recycling sector by elbowing out unregulated practices. This has always been done, as we have seen in other sectors, since the beginning of capitalism, with their all-time co-conspirator, the state. They do not operate under the logic of the market. It is not the invisible hand of the market; rather there is the state handout and monopolization. With their quest for a just and less polluted planet, environmentalists have fallen into the same sinking hole by holding hands with corporations.

Welcome to Sodom (Weigensamer and Krönes 2018), a documentary released in 2018 that won several awards, highlights many of the issues at play in electronic waste dumping and scrapping in the Global South. It engaged with pyropolitics at the famous Agbogbloshie electronic waste scrap yard in Ghana by showing fire, smoke, and black bodies covered with sweat, mining copper, and other metals from mountains of e-waste (Little 2021). Agbogbloshie then became the symbol of e-waste pollution and violence in the Global South. The notoriety and fame of Agbogbloshie influenced governments, development agencies, and nongovernmental organizations, and Agbogbloshie went on to become a global e-waste laboratory for e-waste researchers.

With goodwill, the German government offered a hand to help Ghana with the Agbogbloshie e-waste graveyard problem (Africa Reloaded 2022). In 2017, a 25 million Euro grant was signed, with five million Euros directed toward a technical cooperation program provided by the Deutsche Gesellschaft für Internationale Zusammenarbeit (GIZ), and the remaining 20 million Euros to be used as seed funding for the future recycling fund which came from the German Kreditanstalt für Wiederaufbau Bank (KFW) (Germany Embassy Accra 2017).

GIZ's website outlines the project's goal as follows: "To ensure the Ghanaian Government effectively establishes an environmentally sustainable and socially responsible system for managing discarded electrical and electronic items." In order to accomplish this goal, three objectives are outlined. First, to provide guidance to the Ghanaian Ministry of the Environment and Environmental Protection Agency regarding the establishment and funding of a highly efficient e-waste management system. Furthermore, and crucial to the electronic discards discussion, to advocate for the private sector's interests by assisting enterprises in creating and executing innovative and financially sustainable products and services. Finally, to facilitate the transformation of unregistered trash dumping facilities into ecologically sustainable recycling centers and to incorporate the informal sector into the country's e-waste management system (GIZ 2023).

The last two approaches precisely illustrate what I am bringing forth in this book. After labeling the workers at Agbogbloshie as agents of pollution, the governments of Germany and Ghana are putting mechanisms in place for capital accumulation. In July 2021, the scrapyard was demolished in a warlike engagement between state forces where human rights violations occurred with no reports from international media. The site was closed, leaving workers with no place to work. Consequently, e-waste activities have moved to other parts of Accra. At the same time, the former Agbogbloshie scrapyard has turned into an illegal dumpsite (ChannelOne TV 2024).

This initiative and many others focus on the apocalyptic aspects of electronic discards recycling in the global(s). Rare ingenuities and innovations of workers like those at Agbogbloshie come to light, while refusing to consider the ways in which people pursue their freedom and independence, resisting neo-colonialism through their activities in these spaces. The refusal is not by accident. It is the continuation of the capitalist story that is meant to show only one side of unregulated electronic discards. As I will show in the next chapter, studies conducted at formal electronic recycling facilities in Sweden and Spain reveal exposure to hazardous chemicals to workers and the environment. Any single story, on many occasions, is a worlding of power. In electronic discards, it means resettling the capitalist for accumulation.

Both primitive and salvage capitalism accumulate capital from these disparate practices and places. This presents a double-edged sword for those living downstream of the capital. Primitive accumulation dispossesses peasants' lands and livelihoods, hence pushing them to cities as cheap labor. Salvage accumulation happens by exploiting their relations outside the capitalist system, like kinship and race, for capital gains. As a system, capitalism is characterized by tensions creating and exploiting crises for further accumulation. Techno-capitalism is no exception. It is a technological amplification of both primitive and salvage accumulation. However, I see spaces like Mahakama Ya Friji and repairing acts in Tanzania as a resistance to techno-capitalism.

Techno-capitalism has created a crisis of overconsumption that is leading to several other crises, including those related to energy provision, discards disposal, and environmental pollution. In recent years, the idea of a circular economy has been embraced in techno-capitalism and is peddled as a solution to these crises. However, the integration of the circular economy into techno-capitalism is based on the same principle: profit generation for capital. Rather than embracing the circular economy as the only solution, the world could also benefit from examining the logic and practices of unregulated workers and the attitudes of Tanzanian consumers toward objects and their place in the world. In many cases, these attitudes reflect a deep respect for the environment and a recognition of the limits of material goods. These insights, combined with new technologies and innovations, could help create possibilities for more sustainable and equitable economic systems that benefit humans and non-humans.

E-GATHERING: LOCATING, COLLECTING, AND GATHERINGS

As the title, *Gathering Electronic Waste in Tanzania*, suggests, throughout the book I engage with the concept of gathering. The gathering concept in this book is deeply influenced by recent works of Ursula Le Guin and Pierre du Plessis on the usefulness of the concept to understand contemporary phenomena. In *The Carrier Bag Theory of Fiction*, Ursula Le Guin proposes using the metaphor of the gatherer's bag or container rather than the spear of the hunter to shape the stories we tell. This move centers collectors and their collectives, allowing for different kinds of explorations of labor relations and the relationships that materials have with the landscape (Le Guin, Haraway, and Lee 2019). Le Guin's theory helps us to understand e-discards collectors both through their collective relations and through their containers to comprehend their entangled connections to a multiplicity of material values, flows (where they come from), and their embedded toxicities. We can also understand something about the collectors' labor relations, tools, knowledge, and connection to techno-capitalism through assemblages of these relations.

This book extends Le Guin's gathering concept to e-discards and the social, and material networks in which they are embedded as they travel, and undergo valuation, devaluation, and valorization in Dar es Salaam cityscapes. Gathering e-discards is about assembling value through unmaking and remaking. Pierre du Plessis, in his work among Kalahari hunter-gatherers, proposes the carrier bag theory of landscapes to explore "the heterogeneity of gatherings with which lifeworlds are entangled" (du Plessis 2018, 161). Du Plessis divides gatherings into two perspectives: first, the practice of collecting or gathering things, and second, the act of coming together. Similarly, this book adopts the gathering analytics in these two ways to bring together the actual practices of collecting e-discards and the collective relations of which both the collectors and the materials become part. Further, the gathering is reflected in the different terms collectors at Mahakama Ya Friji use to describe their labor: *kuokota* and *kukusanya/kukusanyika,* which translate to picking or collecting and a gathering or collection in English.

I asked the gatherers at Mahakama Ya Friji how they would describe their labor. The answer was "*ninaokota skrepa/chuma chakavu*" (I am picking or gathering scrap metal). Rama, one of the gatherers I extensively worked with, told me that he would describe his work of gathering scraps from the city as "kuokota chuma chakavu." The common name for scrap gatherers as known in Dar es Salaam is `waokota chuma chakavu/skrepa`. *Kuokota,* which means locating, tracking, and picking, is a common verb that refers to activities involving picking scattered things. Phrases like "*kuokota matunda*' (picking fruits) and "*kuokota kuni*" (picking firewood) are common in Swahili. The verb kuokota is commonly used for scattered objects, not those ordered or privately owned. I use this kind of gathering to track how e-discards workers locate, collect, and transform value. This side of gathering is about knowing the cityscapes, its patches, and its manifestations: where to start, where to go, at what time, and collecting what. The book follows collectors through the city as they evaluate and collect defunct electronics.

Locating, tracking, and picking scraps, "kuokota skrepa" is the first meaning of the e-gathering concept. These are skills that e-gatherers must acquire. E-gathering is not just about gathering defunct devices. Gatherers know what signs to look for, know the cityscapes, and evaluate accordingly. As peripatetic value makers, gatherers must know where to go and what is to be found there (Kelly 2013; Woodburn 1982).

However, when gatherers bring different scrap materials together, it means *wanakusanya* (gathering). At the moment different groups salvage materials, that is "*kusanyiko*" (a gathering), a noun from the verb kukusanyika. "*kukusanya skrepa*" is when scraps are brought to Mahakama Ya Friji from different areas of the city. Upon arrival at the workshop, gatherers, scrap dealers, crafters, and repairers "*wanakusanyika*" (gather) looking for the material they need. Electronic discards gathering gathers relations at the workshop and connects transnational businesses and capital.

The second meaning of gathering, as I have used in this book, is the coming together of material objects, social and labor relations, the market, and traditional techniques. While *kukusanya* means to collect, gather, or assemble, the term "*kukusanyika*" (gathering), as used in Swahili, means the coming together of networks of relations. This second gathering lens focuses on how electronics and discards enact collectivities, gather social-material assemblages locally and extend globally through the circulation of technologies, discards, and toxicity. For example, chapter 4 follows gatherers through the city, connecting their labor to global prices of metals such as copper. It examines the ways they navigate global copper price fluctuations, while chapter 6 describes how charcoal cookstoves, salvaged from refrigerators and air conditioners' aluminum, enhance cooking practices and conviviality in the city. These two chapters show that electronics and discards gather different practices, assemblages, and labor that occur in disparate places.

The book also engages with gathering as a methodological tool to study and understand a city in the global south(s). Dar es Salaam, like many cities in Africa, is a conjunction of informal congregations that makes it hard to grasp its social and economic fabrics. On the one hand, the setting of Dar es Salaam can be characterized by chaotic, proximate bodies of individuals in movement, buildings, cars, businesses, transactions, consuming and discarding, deteriorated infrastructures (busted water pipes, potholes, power cuts), loud music and advertisements, honking—these characteristics depict Mwanayamala, where the workshop is located. On the other hand, there is a modern smart city with high-rise buildings, 5G internet, and "clean" and well-maintained infrastructures, just like downtown Dar es Salaam, five kilometers away from the workshop.

However, when sifted through the city with its polarization, networks of relations between the "two cities" in a city are open to working in incredible ways. The anarchic environment entails "circumvention of domination and the keeping open of many different trajectories of what life could be all at once" (Simone 2004, 18). Dar es Salaam comprises ruins, continued ruination, and "modernity" all at once. Relations are flexible, fluid, and mobile, with complex intersections and entanglements of objects, spaces, and practices. Creativity and constant improvisation are needed while identity, belonging, violence, expulsion, and camouflage become part of the city's lifeworlds. The book takes Mahakama Ya Friji as an entry to understand the city's entanglements. As I walk with discards gatherers around the city to collect discards materials, the city's landscapes, infrastructures, histories, and class relations reveal themselves. The connection between consumers, discards workers and cooking dinner at home is connected to this web of relations.

Building on the carrier bag approach, I suggest that practices of e-discards salvaging might be more fairly understood and better articulated as a technospheric form of gathering or e-gathering. This is the case both in terms of how workers at Mahakama Ya Friji locate, collect, and transform value from discards, and in terms of how these practices enact new collectivities (which, however, are intriguingly rooted in older collectives) that gather social-material assemblages, which propagate worldwide through the distribution of technology, waste, and pollutants. This also applies to new local forms of techno- and material-semiotic entanglements. Building on Le Guin's carrier bag theory of fiction and du Plessis's carrier bag theory of landscapes, Mahakama Ya Friji is the "bag" of the e-discards at the downstream of techno-capitalism that I am peering into to understand and explore discards assemblages, labor, and life in Dar es Salaam.

The story of the dangers of unregulated e-discards activities in African cities has been told. I use the carrier bag theory and Mahakama Ya Friji as the discards container to tell other stories of labor and life downstream of techno-capitalism in which collectivities emerge and are remade through innovative practices surrounded by discarded materials. The stories unwrap how discards are assembled, how they connect to wider value networks in the city, and how e-gathering maps reveal the city's relationships to discards. As a container, Mahakama Ya Friji also gathers society's material entanglements, national histories, and politics, together with global markets.

Global south(s) cities are often narrated to look different from the idealized proper city, the planned city with paved streets, shopping malls, and well-functioning infrastructures. However, where infrastructures are absent, nonfunctional, or broken down, quotidian activities of the urban dwellers fill the vacuum by inserting localized knowledge and opting to become infrastructures for life to continue. These unregulated economies and people as infrastructure come to constitute the city. The city is a de(re)composition of seemingly infinite possibilities of remaking (Simone 2004).

During these entanglements, creativity adopts surprising forms. Microscale, ingenious activities, like the ones I describe in this book, developing around those problems, are usually overlooked by governments, tech companies, and even researchers. These unrecognized activities, which spontaneously arise out of these unregulated patches, create platforms for potential un-projected urbanization; they do not necessarily follow any particular trajectory of urbanization. They are more than just compensational infrastructure for missing social orders and lack of successful urbanization (Simone 2004). African cities must be considered in their complexities, still shaped by colonial and postcolonial ramifications.

As collectors gather material at the workshop, they also gather memories of colonialism, post-colonialism, realities, and imaginaries of techno-capitalism, the city, and the world. When they dismantle and separate different materials from discards, they also dismantle and separate the old world. And when they craft new objects, they become crafters, crafting the new city and the new world from the rubbles of the old world. Their innovative thinking and skills turn ruins and rubble into a new world. Scrap dealers are the intersection between the workshop and the world, a world that is so thirsty for raw materials—a thirst that is drying the planet. They are slowing the overheating of the earth. They are at the forefront of remarking the new world. All that is needed is to stretch our gaze to this new world unfolding.

OVERVIEW OF THE BOOK

The book is organized as follows. Chapter 1 describes and engages with e-discards/e-waste perspectives and discussions. This is a conceptual chapter that aims to put e-discards discussions in perspective with the Tanzanian environment. The chapter builds the foundation for the empirical chapters that follow.

Chapter 2 provides a brief history of informality in Tanzania. It follows historical events from colonialism through independence and the subsequent *Ujamaa* (socialist) period to the liberalization of the Tanzanian economy in the late 1980s. In addition, the chapter explores political changes by examining policy directions toward informality by successive presidents of Tanzania and their policies toward informality. Tanzania has had six presidents since its independence in 1961. Each president promoted different policies toward the informal sector. The chapter lays down a general understanding of the informal sector and its historical friction with the state.

Chapter 3 explores how electronics as technological objects do not hit the garbage can on the first instance of breakdown. There is a gap between importation and when things become defunct. Before electronics are discarded, they make several trips to repairers before gatherers collect them. The chapter juxtaposes the concepts of consumption and motivations for repair in Tanzania. It critically examines repair in Tanzania through the concepts of consumption and care.

Chapter 4 sits down at the Mahakama Ya Friji workshop. The chapter explores gathering as the coming together of labor and material to generate new objects. History, kinship, conflicts, and specializations influence the gatherings. In many media portrayals and the public imagination, informal recyclers have been stripped of their agency, authorship, and voice. This section elucidates how labor is organized and how different tools are employed in performing labor. *Tools* are the essential lenses through which the politics of e-discards are engaged. Workers are viewed only as instigators of pollution in the local and international media and rarely are their labor or life stories at the center of analysis. Tools determine which practices are proper in electronic recycling. Tools are also connectors of workers' imagination and skills that, in return, generate value and produce new objects. In addition, the chapter extends to the labor of the more-than-human in waste labor.

Chapter 5 turns attention to the labor and knowledge of scrap collectors as they collect, valorize, and learn about value. Walking with collectors, gathering, and evaluating discards are woven into and aligned with the history of the city and the class composition of its neighborhoods. This chapter addresses the first part of gathering: locating, tracking, and collecting. Gathering scraps is a skill that has to be learned and mastered. The chapter describes the knowledge about the cityscape and the creative ways that collectors navigate, locate, and gather value from e-discards. The chapter also engages with the scrap business, including how the scrap workers relate to each other at Mahakama Ya Friji and how the scrap business responds to local politics and market forces. Changes in the global market and local politics affect the scrap price; hence, the scrap business is imbricated, reflecting, and creating global value cycles.

Chapter 6 explores the tools that connect crafting at Mahakama Ya Friji. The chapter follows how different objects are produced and the material consciousness that crafters possess. It brings specializations in production and the ways these specializations arise. The chapter also engages with learning crafting through apprenticeship. Additionally, it aims to show how tools are acquired and their effects on power and politics at the workshop.

Chapter 7 follows one of the products repurposed from e-discards. Charcoal cookstoves are central to the cooking and sociality of urban dwellers in Tanzania. They assemble distinct encounters from households to leisure places—charcoal stoves are central to conviviality and are a multimillion-dollar business. This chapter travels through the city, exploring how charcoal stoves enhance social life. In addition, the chapter connects the charcoal stoves' agency to the deforestation of Tanzania's forests.

Finally, chapter 8 analyzes exposure to hazardous chemicals. Toxicity is a concern in the local social world, and this chapter documents how workers at the workshop address the problem. Using the concept of lifescaping (Edelstein 2003; Ntapanta 2021), the chapter observes what workers at Mahakama Ya Friji do with knowledge about exposure when they hold no power to lessen that exposure. Workers employ individual initiatives in their logic to minimize or escape exposure. The chapter connects to society's dilemma and anxiety about the increase of noncommunicable diseases. The anxiety is directed at the modern lifestyle in which electronics are entangled. These lifescaping initiatives are essential for understanding life and toxicity downstream of capitalism.

The conclusion returns to the workshop in 2023. Many events have happened, the losses and imaginations that some have come into reality and some never materialize. Nevertheless, for many who are still at the workshop, discards still offers an avenue to move out of downstream.

NOTE

1. Because some of my walimu passed away, migrated, or changed jobs, therefore, several are anonymized in chapter 3, chapter 4, chapter 5, chapter 6, chapter 7, and the conclusion.

Chapter 1

Electronic Discards

Worlding, Flows, and Property Relations

DISCARDS FLOWS

Sunset in Stone Town, the old town of the Zanzibar archipelago, is always magnificent, except during the rainy season when the weather can be violent following monsoon winds. The evening show has started at the Forodhani food market. Groups of young males dive in spectacular style into the ocean. Forodhani is a beautiful sea and traditional Zanzibar food market. It is built on a concrete bank with enough depth for the divers and scenic sunsets. Food stalls are installed at the center of the square; leisurely people come in for evening snacks. Aromas of fried fish, *mishikaki* (skewers), and Zanzibar pizza cooked with local spices engulf the market. About 300 m away, a large container ship is docking at the harbor, sending waves toward the walls of the food market, pushing the divers in all directions. The boys diving from the cliff are thrilled by the ship's big waves. For us spectators, thinking how dangerous it might be if they are caught in those big waves is scary.

This is the Zanzibar (and Tanzania) entry port for goods from abroad. I assume that electronics also arrive in containers arranged over the ship deck. Those electronics will eventually come out of the port, onto the streets of Zanzibar, and probably to the mainland, making life thrilling and enjoyable for those who acquire them. The devices will save lives, make life comfortable, and allow people to be in the "contemporary" world. But, in fact, sooner or later, those electronics will have failures that lead them to repairers, not only once but probably many times. Sooner or later, they will become defunct and end up in the hands of recyclers.

The Zanzibar port is famous for the availability of cheap used products from the Middle and Far East, Europe, and North America. The port has been popular since Sultan Sayyid Said moved his capital from Oman to Zanzibar in

1840 and throughout colonialism (Wilkinson 1996). After independence and during the Ujamaa period, the port mainly exported cash crops to industrial nations. However, it gained popularity after the fall of socialism and the adoption of liberalization in the early 1990s. Zanzibaris of Arab and Asian descent have had relatives abroad, especially in Oman, the United Arab Emirates (UAE), and the United Kingdom. Many Arabs and Asians fled persecution in the wake of the 1964 Zanzibar revolution (Keshodkar 2014). With liberalization, doors were open for business. Most electronic shop owners in the Darajani market have business connections with these diaspora communities.

A secondhand TV retailer in Kinondoni explained that all his TVs come from the Middle East. His relatives in Dubai collect the devices and send containers through Zanzibar. It is easier to receive containers through Zanzibar than through Dar es Salaam because there is less cargo traffic, and customs and import regulations are more flexible in Zanzibar. I also visited another secondhand electronics shop to buy a microwave and briefly chatted with the shop lady. She also pointed out that her devices arrive through the Zanzibar port.

These two secondhand retailers provide insight into Zanzibar's importance in importing used electronics. Even though laws and regulations are meant to control business between Zanzibar and mainland Tanzania, the loopholes are too big, to the extent that it is quotidian to import used electronics through Zanzibar (Keshodkar 2014). Zanzibar is renowned as a market for secondhand electronic products and has more flexible policies regulating the importation or disposal of goods. There is no regulation of the quality or quantity of imports compared to mainland Tanzania. In Zanzibar, e-discards are treated the same as other forms of waste, opening a legislative loophole for unregulated activities. The main initiative to regulate e-discards in Zanzibar is through the Zanzibar Bureau of Standards, which has a special committee to monitor imported electrical and electronic equipment (Jumbe, Issa, and Carpenter 2014).

The 1990s was a critical time for used electronics imports to Tanzania. During this period, electronics production skyrocketed worldwide, with outsourcing to cheaper regions, mainly in Asia, and planned obsolescence taking a central role in design (ILO 2014). At the same time, the world had started to become aware of the accumulation of e-discards. However, Tanzanians living abroad saw a business opportunity in used electronics.

The thirst for certain imported products had come out of coloniality relations and many years of socialist policies that controlled people's consumption. On the eve of liberalization, used goods became a profitable business (Hansen 2000; Koponen 1986). Though there has been an increase in new products in the market, to this day, lower- and lower-middle-class Tanzanians still prefer certain secondhand products from abroad. Electronics, kitchenware, clothes, and shoes are popular used goods that come through Zanzibar. After arriving at the Zanzibar port, these goods cross to mainland Tanzania, where the demand is higher than in Zanzibar. However, not all products are destined to stay in Tanzania. The ports of Zanzibar and Dar es Salaam, the largest ports in Tanzania, are entrance gates to the whole Great Lakes region.

A secondhand electronic importer to Tanzania calls discarded electronics in the United Kingdom as *makinikia*, a Swahili word that became famous in 2017. The fame can be traced back to President John Pombe Magufuli, who ascended to office in 2015. During his time in office, Magufuli started a legal dispute with the Canadian mining conglomerate Acacia Mining Plc over exporting sand concentrates from Tanzania's gold mines. Apart from gold, the sand was believed to contain other minerals that Acacia was not licensed to mine in Tanzania. Supposedly, the sand was illegally exported out of the country to further smelt the metals. Therefore, Makinikia is the Swahili term for the sand from the gold mines that is believed to hold other minerals.

Therefore, the importer uses the term makinikia to refer to secondhand electronics and other machinery. Even though the devices have been used in England, there is still value or function remaining in them that can be used in Tanzania. Following the Basel Convention that bans transboundary export of hazardous materials, including e-discards, reveals the gray side or probable illegality of the export of secondhand electronics from the United Kingdom to Tanzania. Many Tanzanians and non-Tanzanians abroad, especially in Europe, Asia, and the Americas, collect and export used goods to Tanzania. Tanzania's market for used goods is very stable, making the business lucrative. In that sense, the importer uses the word to mean that even though the objects have been thrown out in the United Kingdom, they still have great value and can be used in Tanzania. However, it also breaks the ban on transboundary movements of hazardous materials.

The Basel Convention, signed in 1989, aims to safeguard developing countries from being used as dumping grounds for hazardous waste. The phenomenon can be described as "pollution colonialism," where developed countries turn less developed ones into junkyards (Liboiron 2021). The convention prohibits the transboundary movement of hazardous materials, including e-discards. E-discards are included because they contain hazardous chemical compounds (Andrews 2009).

Other multilateral agreements and mechanisms are in place to control transboundary movements of hazardous materials, including e-discards. The 1989 Basel Convention is the main agreement, although others include the Stockholm and Bamako Conventions. Even though there are discussions about when devices are waste, countries have put infrastructure and frameworks in place to regulate, control, and manage it.

Sustainable Development Goal 17 under "technology" promotes the development, transfer, dissemination, and diffusion of environmentally sound technologies to developing countries on favorable terms, including concessional and preferential terms, as mutually agreed. However, the fast growth of information and communication technology (ICT) has led to flows of secondhand and nearly end-of-life electronics into developing countries that lack the capacity to handle hazardous materials (Magashi 2011).

In response to Article 11 of the Basel Convention, African countries must prevent the flow of hazardous substances into their countries. Otherwise, they will be victims of pollution colonialism. Under the Bamako Convention, which came into force in 1998, African countries agreed to ban the import or incineration of hazardous waste and not use oceans and inland waters to dump hazardous waste, including radioactive waste.

In response to the surging e-discard problem, the Tanzanian government and the United Nations University conducted an e-discard statistics analysis for mainland Tanzania (E-waste Monitor 2019). The Tanzanian e-waste report of 2019 was the first to describe the importation of e-discards in the country. The report gathered data collected from authorities that directly relate to electronics. The institutions included the Tanzania Revenue Authority (TRA), the Tanzania Communications Regulatory Authority, and the National Bureau of Statistics. The main issue pointed out in the report is the need for more specific policies for e-discard operations (NBS and UNU 2019).

THE DISCARDS PROBLEM

Today, the number and variety of electronic devices in circulation are at once mundane and shockingly large, as homes, workplaces, and everyday life depend on electronic gadgets. Plug-in toasters and kettles are used to prepare breakfast; offices are organized around computers and printers; batteries power cars; and meetings are arranged through teleconferences, emails, or short message services. Electronics are a marker of progress and symbolize human control over time, distance, space, and effectiveness. More importantly, they have become an essential part of ourselves, perhaps to the extent that they may be approached as cyborg prostheses (D. Haraway 1991). However, these electronics have limited lifespans, and with them, waste streams are emerging that affect the planet and regions of the world in uneven ways (Reno 2016).

Like other types of waste, e-discards are also subjective value concepts that are difficult to define objectively. They are not static; materials constantly engage in an ongoing social reconfiguration and reevaluation that creates ambiguity regarding where, to whom, and when they are valuable or worthless objects (Drackner 2005; Gregson et al. 2010; Reno 2009). Constant flows of new electronic technologies and products, together with improvements through more unique and better models, marketing strategies, and people's desires to have better and more modern technology, have encouraged people to continue upgrading to new gadgets and discard older ones.

Several scholars have noted contributing factors to the surging e-discards problem, including planned obsolescence (Wang and Gaustad 2012), which pushes electronics prices lower in the market (Lehmacher 2017); the high cost of repairs (Camacho-Otero, Pettersen, and Boks 2017); increased electrification of nonelectrical equipment and objects, like children's toys, games, and more, with shorter life cycles (Baldé et al. 2017); and the contemporary culture of consumption (Bauman 2013; Reno 2009). In 2019, global e-discards production reached an astounding 52.6 million tonnes; in 2020, it rose to 55.5 million metric tonnes and 62 million tonnes in 2022. A new report from the Global E-waste Monitor indicates that e-discards generation is escalating at a worrying pace, five times faster than documented recycling efforts. This upward trend is expected to persist, with e-discards production anticipated to reach 82 million tonnes within the coming six years (Murthy and Ramakrishna 2022; Statista 2021; The Global E-waste Monitor 2024).

Contemporary consumption patterns are entangled in what Veblen and Lekachman (Veblen and Lekachman 1994) refer to as "conspicuous consumption." Consumption is intertwined with a constant need to acquire new products and with increased speed in waste production. Consumption has also been accelerated by the availability of a wide variety of goods in the market, to the extent that certain values, like repairing, are pushed to the margins of consumer considerations (Dumont 2013). In Norway, for example, repair has been disembedded from spheres of consumption (Ntapanta and Medaas 2024). This means consumers care less or have no agency to repair the electronics when buying them. The tech industry has separated and left consumers powerless to repair devices they own. In many instances, the repair cost is higher in Norway than buying a new product.

Electronics are made to discourage Do It Yourself (DIY) repairing. There are fewer places to repair as the government maintains infrastructures like taxation and tight regulations that kick small third-party repairers out of business. These factors have increased consumers' excitement and need to acquire newer gadgets, which systematically alienates them from the very products they buy (Bauman 2013; Miller 2012; Reno 2009). As more electronics are bought and discarded, they must go *away*, not into "our" backyards (Doherty 2021). They must end up somewhere, and in many cases, in the "wastelands" (Chalfin 2019) or the domiciles of "wasted lives" (Bauman 2013). This means e-discards often end up in places situated downstream of the capitalist system, like Tanzania.

While electronic devices might hold no or low value in certain parts of the world, they emerge in remarkable ways and enter into new assemblages of relations in the Global South(s). In most African countries, low purchasing power, the desire for "modernity," that is, being in the contemporary techno-world, lack of appropriate policies and regulations, and incapacitated and overworked regulatory authorities allow defunct electronic items to get dumped, donated, or resold (Ntapanta 2023). The origins of e-discards arriving in the Global South(s) are often traceable. For example, several computers found in the Agbogbloshie district in Ghana had labels from firms like Jyske Bank (in Denmark), Exxon, and U.S. government departments (Dannoritzer 2010). A survey at one site in Nigeria found that many broken computers had company and government tags, mainly from the United Kingdom, the United States, Germany, Belgium, the Netherlands, Finland, Norway, and Italy (Ewastemonitor 2019).

Upon arriving at these spaces, e-discards encounter other assemblages of value. They become things with value in place. Electronics possess a material agency that is continually negotiated with people who encounter them in different spaces and times, even when they have become obsolete. Their characteristics influence the actions taken upon them and vice versa. Here, encountering e-discards entails everyday negotiations between people and things engaging with each other. During these encounters, new "cultural meanings, categories, objects, and identities" are created and re-created (Faier and Rofel 2014, 363). During encounters, improvised actions and reciprocity of responses produce room for short- and long-term negotiations and the reconfiguration of actors' characteristics.

Electronic devices possess attributes that allow them to be easily transformed and salvaged, even when obsolete (Law 1999; Uden and Francis 2011). Their materiality accepts absorption in the sociality encountered, and their value varies upon entering different networks of relations. Electronic devices offer and exert "flexibility, adaptability and influence" with other actors (de Laet and Mol 2000; Uden and Francis 2011, 26). E-discards materiality demonstrates fluidity by allowing salvaging and transposition into new values, perhaps no longer in its initial design or purpose, but as emergent, fluid, with new uses and potentialities.

The Global North(s) in its collective, while I acknowledge the dynamics within the Global North, by discarding materials and dispatching them to the South(s), establishes certain relationships. Waste to one person is not the same to another—or between communities, societies, or countries. Infrastructure available in developed countries creates particular relationships to what is discarded, enabling discards to be valuated and removed out of sight easily and quickly. However, other relationships emerge in cities such as Dar es Salaam, where waste management infrastructures are in constant shambles and cannot swiftly move out of sight, thus disturbing the order of life and things. Unregulated recycling activities like those at Mahakama Ya Friji step in to fill these gaps.

WORLDING ELECTRONIC DISCARDS

Waste is a product of "worlding," a social and relational concept. It depends on people's relationships with objects that are refused and with each other during social encounters and assemblages. That is the reason for adopting the term "discards," as it provides a broader terrain to think of our refuse. My approach to unregulated e-discards recycling draws on Gayatri Chakravorty Spivak's (1985) use of the word "worlding" to describe the "otherness" of places, people, genders, sexes, political systems, and cultures. This categorical hierarchy is convenient for signifying power over "others" (Raja 2019). Spivak's "worlding" provides a vital tool for thinking about unregulated electronics recycling.

On September 26, 2019, the AnthroTox project organized a public engagement at Litteraturhuset in Oslo titled "What is my phone doing in Africa?" In my presentation, I discussed the 2018 Basel Action Network report that indicated leakages of e-discards transported to developing countries from EU-selected countries of Austria, Belgium, Denmark, Germany, Hungary, Ireland, Italy, Poland, Spain, and the United Kingdom (BAN 2018). During the question and answers session, one of the participants contested my analysis by arguing that in Norway, there are proper systems for collecting and handling e-discards. There is no way discards from Norway are leaking to developing countries. Norway, of course, was not one of the countries investigated by the study. There was no data on any leakages from Norway, as it was not part of the study. I didn't have an answer at the time because Norway was not part of the report.

This is how worlding happens at different levels, from technological thinking, that "we" have the best technology for recycling, to government policies, that "we" have enacted the state-of-art e-discards policy and laws to citizens, trusting that the system in place is better than elsewhere. It is a common argument I have encountered when discussing e-discards in several EU countries. It was not until six months after my presentation, on March 11, 2020, that the Norwegian state media group, Norske Radio og TV Selskab (NRK), released a documentary called "The Waste Smugglers." The documentary shed light on the alarming quantity of electronic discards being illicitly transported in vans from the bustling Norwegian port of Bergen to West Africa (NRK 2020).

The worldings of e-discards recycling—best practices vs. the "burning" practices—are a categorical power hierarchy in e-discards politics, regulations, and discussions. Recycling practices in developed countries are categorically placed and viewed as best practices, while those in developing countries are deemed savagery. The Sodom narrative, borrowed from the documentary "Welcome to Sodom," incites politics over the danger or toxicity of unregulated e-discards recycling in developing countries. At the same time, recycling mechanisms in developed countries are rhetorically framed as the best ways to manage hazardous chemicals, as the Norwegian person during the AnthroTox public outreach.

Those who work in unregulated electronics recycling are classified as a danger, not only to themselves but also to the environment and to others. Narratives and hymns, sometimes using biblical apocalyptic terms, echoed by environmental NGOs and international news media are often louder than the value of unregulated recyclers. Greenpeace International, Basel Action Network, and *The Guardian* are among the frontline singers. Thinking with e-discards narratives, a clear worlding is purposely inserted to consolidate power and control over others by creating a purity-and-danger dichotomy around knowledge, labor, the environment, and morality that ultimately perpetuates capital accumulation (Douglas 2003; Yanagisako 2003).

E-discards recycling numbers worldwide propose that none of the world's regions are doing better than any other. Suppose we juxtapose the amount of e-discards produced by each region against the amounts "properly"—if there is a proper way—recycled. In 2019, the world produced 53.6 million metric tons (Mt) of e-discards, a 21 percent increase in five years. Asia led by creating 24 million metric tons, followed by the Americas with 13.1 million metric tons, Europe with 12 million metric tons, Africa with 2 million metric tons, and Oceania with 0.7 million metric tons. However, only 17.4 percent of the total produced amount, around 9.3 million metric tons, was correctly collected and recycled. Europe led by "properly" recycling 42 percent, followed by Asia at 11.7 percent, the Americas at 9.4 percent, Oceania at 8.8 percent, and Africa at 0.9 percent (Forti et al. 2020). Seven to 20 percent of the amount produced in 2019 is estimated to have been exported for secondhand use, and 82.6 percent, equal to 44.3 million metric tons, was not documented (Murthy and Ramakrishna 2022).

In that case, we find out that Africa is doing no worse than the rest. However, one might argue that most e-discards in Africa are recycled using rudimentary methods and tools, hence releasing their toxicity. The intention of this book is not to create or cement the symmetries between the West and the South, Global North and South, or developed and developing countries; instead, I want to "defamiliarize" the narrative that is always put forward about "informal e-waste recycling" in the Global South(s) and Africa in particular (Liboiron and Lepawsky 2022). We can learn from the ingenuity and creativity of unregulated recyclers and create a possible future against the backdrop of the mensurate problem of e-discards and toxicity embedded in them.

In every e-discards recycling, regardless of the location, hazardous compounds are released. A study by Balasch et al. 2022) detected high airborne levels of halogenated flame retardants and organophosphates at an electronics dismantling facility in Catalonia, Spain. The study also detected certain levels of exposure to workers (Balasch et al. 2022). Similar findings are presented in a study by Julander et al. 2014), which concluded that e-discards recycling leads to workers' exposure to hazardous chemicals in Sweden. The study found that workers in a formal recycling facility are exposed to multiple hazardous chemicals. The study also found high concentrations of certain hazardous compounds in the air, blood, and urine of the workers (Julander et al. 2014). Similar studies have also found high concentrations of different polybrominated diphenyl ether (PBDE) congeners in air samples from several formal electronics recycling plants (Charles, Groskova, and Cahill 2005; Julander et al. 2005; Pettersson-Julander et al. 2004). These studies show that hazardous compounds are not entirely contained, even in highly advanced recycling facilities.

In the same way, findings from natural scientists within our AnthroTox project suggest higher levels of hazardous chemical compounds related to e-discards in Dar es Salaam. Maja Nipen's study, a chemist within the AnthroTox project, on Persistent Organic Pollutants (POPs) in air and soil suggests relatively high concentrations (both in air and on soil) of certain POPs throughout Dar es Salaam (Nipen et al. 2021). Though these compounds can be released from other sources, they are strongly linked to e-discards because of their customary use in electronics. Another study of e-discards as a source of contaminants in Tanzania's coastal and marine environment by Ane Haarr, another project member, indicates high concentrations of several hazardous compounds from e-discards in aquatic food webs (Haarr et al. 2021).

Mary Douglas suggests that danger, in this case, toxicity, is managed through rituals and symbolic cleansing (Douglas 2003). Politics around e-discards spearheaded by developed countries strengthen Douglas's view. For example, the focus on recycling methods, i.e., chopping and burning, which release embedded hazardous compounds. E-discards recycling practices and facilities in the EU, for example, are juxtaposed as having the proper rituals and as a symbol of control of hazardous chemical compounds in comparison to unregulated valorization in the Global South(s), even though workers and the environment everywhere are also exposed. However, unregulated recycling places, labor, and practices, like at the Mahakama Ya Friji, are the danger.

Working with discards is also translated as disposable labor in many ways; so are the lives of those working with it considered disposable. Discards workers are contaminated by the materials they work with, and they also become also "dirty," which means disposable (Bauman 2013; Doherty 2021; Douglas 2003). Here, "dirt' refers to a more expansive notion of pollution and taboos, meaning that discard collectors become associated with anything unwanted or undesirable by society. I observed and experienced these forms of stigmatization and categorization (see chapter 4). Collectors are called names like crows, equated to drug users, and sometimes accused of thieving. The stigma arises from the sense of being contaminated. Stigma involves a victim identified by an observer and marked according to characteristics that are socially acceptable to stigmatize (Edelstein 2003). The identification of those socially constructed as contaminated characters by an observer leads to negative connotations and discrediting of the victims. Once an identity is stigmatized, the victim is seen as less than human. This means that the whole of waste labor and the identity of the laborer is stigmatized, just like calling Agbogbloshie Sodom or a graveyard (France 24 2019; Weigensamer and Krönes 2018).

In response to these perceptions of unregulated recycling, solutions have focused on neo-liberal technocratic logics that are doomed to fail, for example, attracting investments, building high-tech plants to replace unregulated activities, and formalizing the sector (Little 2022). Regulating discards labor is challenging for Dar es Salaam and Tanzania. Regulating e-discards faces challenges because of its entanglement with livelihood on one side, and embedded hazardous chemical compounds on the other. Regulations for e-discards work well in developed countries like Norway, except when electronics are smuggled from return and recycling centers to be sent to developing countries (NRK 2020). If one were to ask whether these kinds of regulations would work in a Tanzanian context, the answer would be no. Unregulated recycling is embedded into the fabric of the city, the economy, and social-cultural assemblages; regulating with the same approach or infrastructure as in developed countries will not work.

Instead, there is a need to recognize these activities' contribution to e-discards solutions. In developing countries, with the rapid increase in population, the exodus of young people from rural areas to urban centers, partly because of the failure of agriculture—mainly cash crops—and continuously increasing unemployment, more people with no jobs are joining the unregulated recycling industry. Recycling offers a livelihood. The income recyclers receive from their labor cannot be compensated by formal industrial recycling. As I detail in chapters 4 and 5, their income, though fluctuating, depends on the good or bad days; when tabulated, recyclers earn a decent income, sometimes more than those in the formal sector. Hence, unregulated recycling will not be replaced but will operate parallel to the formal one.

Informal arrangements for e-discards collection in Dar es Salaam are also a win-win situation for the discarders and collectors. Collectors buy defunct devices from the users and take them to their workshops. Both the discarder and the collector earn something out of the discard. The formal sector usually charges collection fees. Many people, especially in mid and lower-income areas, prefer to earn, even if it is a small amount, from their discards. They will definitely not pay a fee for their old electronics. In the meantime, economic growth in developing countries, a tremendous decrease in the prices of electronics, the entrance of affordable Chinese products, and changes in technological invention with planned obsolescence, where devices are purposely produced to have a shorter lifespan, produce immense amounts of e-discards, hence more work for the recyclers.

Rather than a source of danger, discards workers have become essential to the economy. Their value-generating activities operate with minimal investment and close gaps that are found in the city's waste management programs. They collect where formal waste collection companies can hardly reach. They mine metals embedded in electronics needed for further manufacturing that would have otherwise ended up in landfills.

PROPERTY AND DISCARDS

Anthropological theories of property and value firmly rely on the sociality of property and property relations. That which can hold value or be acquired as property in one situation might be worthless and unclaimed in other circumstances (i.e., societies, individuals, places, or times) (Turner 2017). We can zoom into micro- and macro-level social relations through anthropological lenses to discover how property, value, and discards operate within fluid boundaries and edges.

There are exciting connections between property, value, and discards. All of them are the product of social environments and are somehow entangled together. Even though property and value can exist independently from discards, there is no way to analyze discards without bringing up value and ownership. The very simplistic definition of discards is an object with no or low value. The object may or may not be the property of an entity, but there must be a relationship for an object to become discards.

When an entity (individual, group, or company) decides to give up an object and throw it away for whatever reason, that entity is alienating itself from the right to own the object. What I mean here, for example, the general logic is that after a municipal waste truck has picked up discards, the entity that owned them has automatically transferred ownership of the discarded objects to the municipal authority. This relationship can be confirmed by strict regulations imposed upon access to discards facilities like incinerators or dumpsites. In other words, if a dumpster diver collects an object from a dumpsite, they obtain a degree of ownership over that object, regardless of whether their collection was legal or illegal. They can use the value for their own gain (Ryan 1988). When discards gatherers pick up objects, they assume ownership over those objects and the same property relations exist. For them, the object is not waste. It is property with value.

This way of thinking about discards can be well articulated through Michael Thompson's "Rubbish Theory" (2017). Thompson argues that an object will keep decreasing in value until it reaches a certain level of devaluation. When an object reaches that level, it becomes rubbish or a discard in that particular time and place. Thompson further argues that an object will continue in a timeless and valueless limbo until it encounters a social environment that exhumes its value (Thompson 2017).

Most people, especially in developed countries, have electronics lying in their basements, attics, or drawers. Some of them are old technologies that they will not use again. Some are there for different reasons, such as being broken. Just like my coffee machines that I mentioned in chapter 3. At this moment, the devices are in a limbo state waiting for a particular social encounter for them to either be discarded or gain value. When discarded in a dumpsite, the classification has already taken place; the device is considered a discard. The same when the device becomes an object antique value, the owner will have a different relationship with it.

The same happens in Tanzania. When gatherers buy and bring discarded electronics from the city to the workshop, they become more than just discards. The same thing also occurs when discarded electronics from abroad arrive in Tanzania: they encounter an environment that alters their value. Tanzania's social and economic environment allows items once regarded as discards in other parts of the world to take on new value through refurbishing, reusing, or repurposing.

At Mahakama Ya Friji, electronic discards are gathered and then salvaged, creating new value in ingenious, creative, and innovative ways. These practices offer "coherences, interfaces, and connections" of value-regenerating activities that support people in tackling daily life challenges and feed back into global value chains (Mavhunga 2017). Hence, electronic discards become more than just "discards" or "disorder." Instead, they enter and converge into new assemblages and networks. Electronic discards take on new values that create livelihoods at local levels and that are surprisingly connected to national and global economies in very specific ways. When electronics are discarded, they provide an avenue for thinking about how, where, and who knows how to resuscitate and harness value out of them.

CONCLUSION

The advent of single-use disposable items coincided with the launch of the initial women's sanitary pads, known as Kotex, which were developed in the 1920s. Kotex marked the inception of disposable, one-time-use items and the emergence of deliberate obsolescence (Strasser 2000). The observations I made in Dar es Salaam are comparable. Previously, individuals and communities worldwide engaged in the practice of repairing, mending, and refurbishing items, which is still the case in Tanzania today.

The rag and bone collectors were frequently gathering discarded items. Blacksmiths and craftsmen fashioned objects using collected materials. The extraction and utilization of value from various things was documented by Strasser in 2000. This underwent a significant transformation throughout the period between the World Wars and the 1950s. The expertise in mending, repairing, and reusing has gradually diminished in the majority of affluent nations. The markets were inundated with inexpensive products that had a limited lifespan and were challenging to fix (Humes 2013; O'Neill 2019; Packard 1969; Rathje and Murphy 2001; Rogers 2006; Rufus and Lawson 2009; Strasser 2000).

In this book, I examine individuals who engage in the activities of acquiring, repairing, collecting, crafting, and trading discarded electronic items. These ubiquitous groups continue to play a vital role in Tanzanian society. The circular economy concept is used as a solution to address the complex waste economics and the worsening waste crisis, serving as a roadmap toward achieving zero waste. While the concepts behind a circular economy are attractive, the actual execution poses significant challenges, particularly on a global scale.

In the Global South, the implementation of the circular economy has primarily relied on the replication of ideas produced in the Global North without significant modifications. This evokes memories of the 1990s when Tanzania implemented the Structural Adjustment Program. Universal mechanisms are destined to fail. By considering localized strategies for managing and adapting to discards, such as those described in this book, we can discover more effective approaches to incorporating the circular economy within these communities.

In this book, I have presented the experiences of workers in resource frontiers who exert significant effort and occasionally subject themselves to various health hazards in order to complete the value cycle. Discard collectors are equally vital, even though conventional economics does not regard gathering as a crucial component of the value cycle (Alexander and Reno 2012; Lepawsky and Mather 2011). Discards gathering occurs on the periphery of resource recovery. An issue with the circular economy model is its lack of incorporation of unregulated discards recovery from the margins. In the regions commonly referred to as the Global South, the involvement of expatriates in various endeavors is mostly driven by international organizations and dominated by the elite class, disregarding or occasionally enlisting unregulated discards workers in projects without adequately considering their agency.

Chapter 2

Informality and Its Malcontents in Tanzania

A Brief History

When I arrived in Dar es Salaam in July 2018, I contacted Salim, a TV repairer, whom I had met two years back while doing my master's degree fieldwork and internship with UN Women Tanzania. At the beginning of the fieldwork, I was still unsure where to start and what objects, problems, or people to follow. Therefore, I had my first interview with him—a general one to understand the electronics ecology in Dar es Salaam. Salim had worked as a marketing officer and a network engineer for MultiChoice, a South African TV cable network. By the time of the interview, he had left the job for self-employment/entrepreneurship (*mjasiriamali*)[1] for about a year. We met at Uhuru Peak bar, where he hung out most evenings. Salim did not have an office; in that case, Uhuru Peak was his office.

It was a bright afternoon, the sun was shining, and Dar es Salaam's heat had lowered slightly. Salim was sitting at his usual table under a tree near a charging cable. His three phones were all connected to power. I remember when we met in 2016, Salim loved American hip-hop outwear. He would have worn his basketball cap, a t-shirt, jeans, nice sneakers, and a shiny watch. This time, he had a completely new look: a long-sleeved shirt, well-shaven hair, casual trousers, and a pair of oxford shoes. Salim was a school teacher back in the day, so I thought maybe the new outlook was refreshing the memories. He fancies himself that he was a fantastic mathematics, physics, chemistry, and biology teacher.

Nevertheless, he quit his teaching job too, to study finance. After completing his bachelor's, he did not find jobs in finance. After a long stretch without a job, he took a marketing position at MultiChoice and later joined the engineering team. It was here where he ended up learning how to install and repair TV networks. "I cannot continue wearing hip-hop outfits. You know I work in people's houses, installing their TVs, some of them are high profile.

I cannot wear like *muhuni* (a hooligan)," he explained about his change in outlook.

Swahili speakers of Tanzania like to use specific terms to point out people's class, jobs, race, ethnicity, and other groups. These terms are going to appear in several vignettes throughout this book. Therefore, muhuni means a hooligan, profligate, wastrel, gadabout, or lawless person (Burton 2005). Someone can become a muhuni in many ways: how one dresses, speaks, or can be a lifestyle. For Salim, being referred to as a muhuni would jeopardize his business. Therefore, he had to change his lifestyle and how he dressed to become a *muungwana* (a person with morals), which gives *heshima* (respectability). Like muhuni, muungwana also earns heshima from outward representation, like how they dress, speak, classiness, and interact with neighbors (Hillewaert 2016).

With his new outlook, Salim also drank tonic soda. "I have stopped drinking," he added. Salim was a prominent member of Uhuru Peak's VIP table. In fact, I met him there when I went for a drink myself. Only a handful of people were allowed to sit at the VIP table. Every evening after working hours, the table was occupied by a group of friends who met there for a drink. Now Salim is not drinking anymore simply because being drunk frequently takes one's heshima away. He has become a "good" Muslim too. At least he attends Friday prayer. Salim had become a respected person, at least an important one. We had to stop several times during our interview so he could take calls. Almost at the end of our conversation, the phone rang again. "I need to call my apprentices. We have a job at Mbezi beach, they have to go now. I will join them later," he warned. While we finalized our conversation, the two apprentices appeared with backpacks full of tools. The two are among the four apprentices that Salim had recruited. Since Tanzania shifted from monologue to digital TV broadcasting, Salim claimed that TV repairing and networking jobs had increased. Before the shift, only antennas and cables were used, and they were easy to connect. With digital decoders and satellite dishes, people with skills are much needed. Salim and the apprentices were busy.

The discussion was helpful to me. I learned a lot about how TVs work, the differences between used and new Chinese imported TV brands, and why people still buy used TVs from Europe. Before Salim left, I asked if we could meet later in the evening. He shortly answered, *"Tutaona mkuu, ujasiriamali mgumu. Nikiwahi nitakushtua"* (let us see, it is difficult when you are mjasiriamali. I will write to you). The phrase mjasariamali comes from two Swahili words: *Jasiri* and *mali.* Jasiri means being brave—someone who takes risks to achieve a particular goal(s). Meanwhile, mali denotes wealth and property. In that case, someone brave enough to start a venture with risks of failing or not achieving the planned goals is called mjasiriamali. Mjasiliamari is a new

Swahili noun coined in the wake of trade liberalization at the beginning of the 1990s. Its English correspondence could be an entrepreneur, in a condensed conceptualization as self-employed (Blanchflower and Oswald 1998). However, who can be regarded as an entrepreneur in Tanzania differs from other parts of the world. In Tanzania, mjasiriamali has little or no capital to launch/succeed in a business, and in many cases operates at the interface of formal and informal grounds.

In December 2018, the government of Tanzania, through the President's office and the Tanzania Communication Regulatory Authority (TCRA), announced a significant formalization project for informal businesses—Wajasiriamali. The Presidents' office announced that all informal businesses had to get *kitambulisho cha ujasiriamali* (entrepreneurship identity card). The President announced that whoever had the card should not be disturbed by the government, and they could conduct their business without being bothered by authorities.

At the same time, the Tanzania Communication Regulatory Authority announced that all electronics repairers should have a license from the authority. The announcement was in accordance with the Electronic and Postal Communications (Licensing) Regulations of 2018. Unlike the kitambulisho cha ujasiriamali issued by the president, the repairing license will be provided upon completion of a vocational training course.

On the topic of the repairing license, two opposing opinions arose during my discussion about the matter with electronics repairers at Uhuru Street, one of the main streets for the electronics business in Kariakoo Market. The first group was pro-license/card and argued that the license/card would give them security to work without fear of authorities. The license would differentiate proper *fundi* (repairers, fixers, or builders) from "*wababaishaji*" (not real fundi). This means a repairer with the license assured customers of their skills, hence attracting more customers than those without the card.

The other group thought the license was primarily aimed at collecting taxes. For years, the government had been chasing unregistered businesses to pay tax. They argued that the license would be used by the authorities to track them down and force them to pay tax. "It is not like we do not want to pay tax, but we do not make enough," said Msomali, a mobile phone repairer. Msomali thought the license would reduce his income, which had been dwindling because of the influx of many repairers in Kariakoo. Msomali had been working in Kariakoo for about five years. He complained about the good old days when there were few repairers. "We had so much work that we had to work from home at night, there are not that many customers these days," he added.

This chapter combines analysis of recent history in relation to the changing political landscape of Tanzania. Selected events, such as the president issuing identity cards and attempts like the TCRA to regulate repairs, provide background and context to situate themes, concepts, and practices described in this book. Places like Mahakama Ya Friji, crafting, discards collection and recycling, repair and knowledges about exposure to harmful chemicals are at the center of colonial histories, postcolonial politics, and, increasingly, postcolonial urban life. This chapter offers insights into why and how the electronic discards problem emerges, assembles specific networks, and contextualizes forms of ingenuity and innovation in recycling workshops like Mahakama Ya Friji.

The chapter reviews policies from colonial times to contemporary government in order to provide an account of Dar es Salaam and unregulated economies to date. The chapter starts with a recent effort by the late president John Pombe Magufuli, who aimed to resolve the problem of unregulated businesses that have been a headache to the Tanzanian government. For decades, the government had failed to regulate and collect revenue from the informal sector. While his government lasted for 1935 days before his sudden death, he reshaped the politics of informality in Tanzania. Electronics recycling workshops like Mahakama Ya Friji are tied to these histories and political processes that shape where and how discards emerge, become entangled, and their gatherings in the cityscapes.

THE BULLDOZER

President John Pombe Magufuli, who came to power in 2015, was described by people as a revolutionary leader, a political idol like Julius Nyerere, a reformer, a Pan-Africanist, and a game changer. He was seen as a fighter for the *wanyonge* (weak)—including wajasiriamali—and a champion against corruption and the rampant misuse of public funds. He jailed, fired, fined, disciplined, and challenged supposedly corrupt public servants and famous businesspeople. He had several nicknames, like *Jiwe* (the Stone), the Bulldozer, and Baba (the father). The popular hashtag #whatwouldMagufulido became a trend on social media and in national and international media. The hashtag emerged after he squeezed unnecessary government expenditures, including a ban on government employees from traveling abroad and the cancellation of the Independence Day celebration. The funds allocated for those events were instead redirected to development projects like making desks for pupils in primary schools. The hashtag meant that before spending money or doing something, one should think, "What would Magufuli do?"

On December 6th, 2016, he directed responsible authorities that wajasiriamali use of the cities' spaces be negotiated, not imposed. Relocation of wajasiriamali will only occur where regional commissioners ensure adequate alternative spaces. All people have equal rights in urban spaces. Therefore, forced relocation is over. Regional and district commissioners are solely responsible for managing street businesses in their regions, and Wajasiriamali have the right to work anywhere in the city but not randomly.

An expected event happened a few months later. The Mwanza City Council decided to relocate street vendors from the city center. The event escalated quickly into violent protests after the city's parapolice and police used extra force to disperse vendors. Magufuli "walked his talk," defending and standing by the weak as he had popularized himself by doing. His reaction was not an accident and was well anticipated based on promises made in his speeches.

President Magufuli condemned the unnecessary force used in the incident in Mwanza and ordered administrative bodies and municipalities in all regions to issue entrepreneurs' identification cards. The president envisioned the card to legalize the informal economy. The identification was a step toward combating unemployment, poverty alleviation, and a solution to the failures of the government to collect revenues from the informal sector.

The new policy of formalizing informal workers through these special identity cards was introduced and implemented through the president's pronouncements. The parliament did not pass any legislation to provide a legal basis for the policy. Regional and district commissioners, appointees of the president, headed up the issuing exercise. The commissioners had to identify, register, and locate the people and businesses as well as the districts and streets in which they were allowed to operate and work. The initial plan involved collaborating with the Tanzania Revenue Authority (TRA) in order to collect taxes. However, by law, the TRA could not collect tax from unregistered traders without legislation passed by the parliament. With such a setback, on December 10th, 2018, the president decided to design, print, and launch the distribution of the ID cards himself. He issued card number T.0000 to himself as an example.

The president authorized 670,000 ID cards. Initially, the target group to be issued the identity cards was street vendors. However, the project was later extended to any unregistered businesses with an annual turnover of less than 4 million TZS (1734 USD). The government collected a fee of 20,000 TZS (9 USD) per annum per business or mjasiriamali. The regional and district commissioners started implementing the ID distribution to the targeted groups with these orders (The Citizen 2018).

A study commissioned by the International Labor Organization pointed out that the high costs of entering and remaining in the formal sector perpetuate either the exit or resistance of small businesses to joining formal arrangements. Projects that tried to formalize the informal sector elsewhere lacked incentives for informal businesses to join. The study found that reforms that reduced the costs of formalization by 50%, as well as the length of formalization procedures, had significant positive results in the numbers of registered enterprises (Elk et al. 2014). The Magufuli project seemed to address these setbacks. First, the IDs were affordable to most informal workers. The process was supposed to be easy and quick. Tedious monthly and daily fees from the local governments were removed, and there was a security incentive to do business anywhere in the registered district without harassment from the cities' parapolice.

Workers at Mahakama Ya Friji had different reactions to the cards, like repairers with the TCRA license. The scrap collectors jumped at the opportunity. The card was important to them in two ways. First, it protected them against harassment from police and sometimes from the public, which is common in their stigmatized profession. Second, they were proud that the government recognized them, and that they could collect from anywhere without fear. The crafters didn't think it was necessary for them to have the card. The reason behind this is that the land where their workshop stands belongs to the ruling party. Hence, they felt secure and did not want to lose their income by paying for the cards.

However, because there were no guidelines on how to go about the distribution of the IDs, the process became a source of political capital for politicians in the ruling party. The implementation focused solely on the number of cards distributed rather than the conditions of the targeted groups. Cases that caused the efforts to backslide started to emerge. In one case, a regional commissioner came out with a list of 47 target-group categories, including small miners, bartenders, taxi drivers, bus conductors, and even freelancers, journalists, and Tanzania National Social Security Fund beneficiaries. The practice stimulated some licensed and tax-paying businesses to join the project. Some companies secretly entered the street vending business by giving merchandise to street vendors to avoid taxes (George and Olan´g 2020). The resulting confusion redirected focus to anyone who could get hold of an opportunity, while the wanyonge who needed secure business environments were left in limbo.

The informal sector is the backbone of the livelihood of millions of Tanzanians. The informal economy is defined as those activities that are not captured in national accounts, operate independently from state monitoring, and involve little or no access to the organized market, credit institutions, formal education, or training. The activities can be legal or illegal but always operate outside the realm of government taxation and regulatory systems (Bagachwa and Naho 1995; Hart 1985; ILO 1991; Maliyamkono and Bagachwa 1990).

Most households in Dar es Salaam depend on informal labor arrangements for their livelihood. The widespread presence of unregulated economic activities in Tanzania, specifically in urban centers, can be traced back to policies enacted by German and British colonial administrations and the governments that followed independence. Early contact between the coast of East Africa and traders from the Middle and Far East marked the beginning of urbanization. Later, colonial policies such as taxation, a monetary system, and the resource extraction economy created blueprints for what would become today's urban centers. For people with no skills or opportunities to fit into these new urban landscapes, informality started to emerge on the margins of colonial towns. After independence in 1961, the adoption of Africanization policies, and then later economic crises and liberalization, have all contributed to the rise and dominance of the informal sector in Tanzania, as I will show in the coming sections.

Over the years, the government of Tanzania has adopted several frameworks and projects to formalize the informal economy. None of these initiatives have produced the desired results, to regulate and collect tax. In this regard, the informal sector has remained the main income channel for many urban dwellers and, at the same time, vulnerable. Lack of access to training or credit, exclusion from decision-making, and lack of legal and social protection are among the challenges the informal sector in Tanzania faces today (Steiler and Nyirenda 2021).

Since Keith Hart (1973) first proposed the formal/informal dichotomy, the concepts have been pivotal in analyzing contemporary African economies. Although Hart's dichotomy fails to capture the complexities of African urban economies, as Hart has commented countlessly, it still offers a framework to analyze surroundings beyond capture, interlocking and fluid livelihood forms and lifeworlds in these complex interlocking cityscapes. Many urban dwellers enter, exit, and re-enter the informal economy. Informality is easy to enter, relies on indigenous resources, is organized around kinship and ethnic affiliations, usually operates on a small-scale, although it is labor-intensive, it has flexible and long working hours, needs to adapt to technological changes, requires skills acquired outside of formal school systems, and is characterized by unregulated and proper competitive markets (Bromley 1978; Hart 1985). As I will describe the Mahakama Ya Friji in the coming chapters, the workshop aligns with these features, as do many unregulated businesses around Dar es Salaam.

THE WORKSHOP IN A POLARIZED CITY

Mahakama Ya Friji is located in Mwanyamala, Kinondoni municipality. A few buildings hide the workshop from passers-by on the busy Kawawa road, just 300 meters away. A small road stretches along the building's front for at least 250 meters, then hits T-shaped junctions in both directions. The road has minimal traffic even though the workshop is in the middle of a bustling neighborhood. Scrap collectors occupy the street in the mornings and evenings without much disturbance from traffic. A secondhand car parts shop on the other side of the road is owned by a mechanic who spends most of his time fixing cars.

Next to the car mechanic's shop stands a wall with a red gate at the front. Behind the wall, there is nothing but barren ground where houses used to stand. The workers told me that a wealthy person bought three houses in the compound and then erected the wall around the area two years ago, but construction has not started yet. However, the discards collectors are happy, as the area in front of the gate is used for burning electrical wires to extract copper. Next to the wall, there is another tall wall topped with barbed wire. No one knew who the owners were, but the wall and the house pointed to the fact that wealthy people lived there.

The workshop occupies a strip of 15-meter wide and almost 30-meterlong. The building is made from iron sheets and is open on both sides. One could look from the front all the way to the backside. There is an open space at the back of the workshop. Several charcoal cookstoves and BBQ stoves are left to dry in the sun. The stoves are painted black, and their interiors are made of clay. They need at least a day in the sun to dry the clay. There are also several shaded extensions to the main workshop on that side. About 10 workers use them as their stations. A small bar with a kitchen is located on the south side of the workshop, with two kiosks on both sides. The district Blind People's Association occupies the building to the left, followed by two food vendors' makeshift stalls, and an evening school attached "ironically" to the district's Occupational Safety and Health Authority office.

Kinondoni, where Mahakama Ya Friji is now located, is a strategic location for this kind of workshop, between residential areas good for scrap collection and industrial ones where scrap can be sold. The neighborhood is at the heart of the city. It is between opulent neighborhoods in the northeastern part of the city, where the wealthy and middle class live. These are areas for collection. It is impossible to set up a workshop there. However, from Kinondoni, it is just a matter of crossing the Bagamoyo Road to reach the area.

Furthermore, to the south, west, and north are mixed areas, uswahilini, and upcoming middle-class areas. Some of these areas stretch up to 30 km from the city center. Collectors cannot walk that far. That is the reason for their concentration in Kinondoni. In the same space, Kinondoni is in proximity to industrial areas, such as the Mikocheni and Pugu Road industrial areas. Here are the markets for metals. Copper, iron, and other metals extracted are gathered here before being repurposed or exported for further manufacturing.

The market for charcoal stoves, the leading products salvaged from scrap materials, is also in Kinondoni Municipality. Stove markets, including the famous ones in front of the American embassy, the Mwananyamala market, and the Kariakoo market, are nearby. For crafters, this is central to their business. First, they get raw materials at a reasonable price. Second, they do not incur additional costs for transporting their products to the market. Furthermore, Kinondoni is also the heart of Dar es Salaam nightlife. Kinondoni never sleeps. There are bars, restaurants, and nightlife spaces on every corner of the district. I will come back later to the relationship between this vibrant urban life and the workshop in chapter 7. The most important to mention is that all bars in Dar es Salaam prioritize food in their spaces. That is where the profit is earned. There is hardly any bar without a kitchen. Nightlife is closely related to all types of cuisine—BBQ, cow legs soup, intestine soups, chicken soup cooked with vegetables, plantain banana, and potatoes, etc. The bars are the essential customers of the charcoal stoves produced at Mahakama Ya Friji. This is why the workshop has to be in Kinondoni—although it has also occupied many different locations across the city, as I describe below.

INFORMALITY AND THE COLONIAL CITY

The workshop's location reflects a long history of geographic and economic divisions in the city. In 1862, fleeing from political and diplomatic pressure in Zanzibar, Sultan Majid Bin Sayyid settled at a small fishing village, Mzizima, and named the place Dar es Salaam, meaning harbor of peace (*Bandar as-salâm*). Majid had experience in long-distance trade and coconut farming, which he brought to Mzizima. The natural harbor of Dar es Salaam invited traders from Hadramaut, southern Yemen, and India to engage in trade and farming. Sultan Sayyid Majid's death in 1870 marked the beginning of the end of Arab rule along the mainland coast (Brennan et al. 2007).

The Anglo-German agreement of 1886 overturned the Sultan's dominance over the mainland coast. German rule placed conditions on the Sultan's collection of customs duties, while his jurisdiction was limited to 10 miles of the coastal strip. Later, Germany occupied the whole coastal line. In 1891, the town of Dar es Salaam officially became the new capital of colonial German East Africa. Looking back at colonial ordinances from Germany, the land and building ordinances claimed to arrange structures for future colonial investment in the city. The ordinances outlined racial divisions through different European, Indian, and African zones and building standards. The Land Ordinance, passed in November 1895, was a critical piece of urban legislation, creating guidelines for determining which peri-urban lands were communally owned and which were ownerless. It also determined ways for Europeans to purchase peri-urban land from local Africans, subject to government approval. *Wajumbe* (local leaders) were the leading actors for authorization;

these transactions gave shape to German-era urbanization in the city (Brennan et al. 2007; Iliffe 1979).

During the nearly three decades of their occupation, until 1916, Germany tried to organize the urban space of Dar es Salaam to suit their colonial project, implementing class divisions and racial segregation. The city's structures and urban planning were designed to protect the welfare of the German rulers. August Leue, town administrator (1887–1890), insisted that the intention was to safeguard the city. Unlike other colonial towns, like Mombasa and Nairobi, the Germans focused on establishing structures and designs that resembled Germany's colonial settlement, that is, wide roads to allow communication from the center to the *askaris* (soldiers) posts. There was a permanent frontier positioned by the harbor and barracks on the land. The askaris' huts were built permanently adjacent to the town *bomas* (fort). The role of the askaris, apart from protecting the city from invasions, was also to monitor and limit Africans' access to the center. So, only those with formal employment, in which most Africans were housemaids and gardeners, were allowed to enter the center (Iliffe 1979).

The city was partitioned into three zones, divided by both race and access to formal employment. The city center, near the harbor and the contemporary Sokoine Drive, was designated for administration and business. Beyond the administrative center, there were *shamba* [field or farm] fields with coconut plantations, and the African villages were located in the remote parts of the city. The Germans kept the Indian and Arab traders who had already invested in the city center. These investments were necessary for German colonial urbanization. The Indian Bazaar had two market halls in the Indian quarter, which stood as the local businesses' focal point. The selling of goods was limited only to Asian merchants. Again, Africans were restricted to shop helpers, dock workers, and housemaids. Africans who did not get jobs in the European or Indian Bazaar were not allowed to enter the two zones (Brennan et al. 2007). Nevertheless, people from the countryside flocked to the city in search of wages.

Between 1912 and 1913, the building ordinance was introduced as a guideline for residential building standards, commonly known as a tripartite division among Africans, Asians (Indians and Arabs), and Europeans. The ordinance introduced limited land transactions in Msimbazi, Msasani, Mtoni, and Kitschwele. The colonial administration purchased Indian and Arab lands in Upanga for European development. The administration acquired Schöller's shamba (formerly owned by the Sultan) for a cordon sanitaire (*Mnazi Mmoja*) and later resettlement of the WWI Carrier Corps. The surrounding area later came to be known as Kariakoo among Africans, a mispronunciation of British colonizers' First World War carrier corps,' whose veterans had been settled there (Brennan, Burton, and Lawi 2007).

After WWI, a new era began under British rule. Nothing much changed; the British adopted most of the German laws. The three zones established by the Germans were officially adopted in 1924. European inhabitants were overwhelmingly located in Zone I, which included the old German quarter, northeast of the city center, and embryonic coastal suburbs to the north, around today's Msasani Peninsula. Zone II was for India's crowded bazaar, which occupied the residential and business quarter, or today's Upanga. Zone III hosted the Africans, with Kariakoo and Ilala as the core settlements. With this separation, Zone I was arranged with trees and lavish vegetation facing a sea view, and Africans' entry to Zone I was restricted unless they were employed by the Europeans. The place was well-known as *Uzunguni* (the European place). On the other hand, Acacia Avenue from Kivukoni to the railway station was a shopping area where Indians owned most of the businesses. Their combination of residences and shops made the area known as *Uhindini* (the Indian place) (Brennan, Burton, and Lawi 2007; Iliffe 1979).

Informal businesses were not allowed in the city center and the Indian Bazaar. In the same way, the British restricted Africans from trespassing in these zones. However, in the African quarters, selling and buying imported merchandise took great shape. The return of Africans who traveled to Europe and Asia to fight in WWI increased the momentum of consuming European goods. Transactions on cigars, Western alcohol, and other products started to appear in the black market among the African populations.

When Tanzania achieved its independence in 1961, President Julius Kambarage Nyerere, the first president of independent Tanganyika and later Tanzania, initiated the African modernization project. Zone restrictions were lifted. However, the Indian Bazaar remained as Uhindini to this day. Uzunguni, Zone I, received new dwellers as the new political class replaced the British. The project also aimed at overhauling the colonial economic structure and bureaucratic arrangements to fit the idea of African modernization. Even though the government took over the economy, the colonial bitterness toward informality was even solidified. Nyerere's government, which lasted from 1961 to 1985, constantly attacked the informal sector.

UJAMAA AND THE RISE OF INFORMAL ECONOMY

During these early years following independence, the economy was still heavily dependent on the colonial infrastructures. At independence, 70% of the bureaucracy was made up of noncitizens. The British divide and rule strategy was still in place by allowing chiefs to run local politics and economies. In 1964, Tanzania adopted its first five-year development plan. The plan targeted private-sector investment, foreign assistance, and diversification of production as vital tools to achieve agricultural transformation, boost manufacturing, and build infrastructures (Yeager 2020). The economy depended on exporting cash crops to feed industries in Europe.

The adoption of Ujamaa (socialism) following the Arusha Declaration in 1967 led the government to take control of the entire economy. First, the government created public parastatals to run essential sectors of the economy. This was followed by the centralization of land and the nationalization of schools, health institutions, banks, insurance companies, and industries (Lugalla 1997). By the early 1970s, Ujamaa ideologies, policies, and administrative arrangements had a deep influence on the economy (Boesen and Koponen 1986). In doing so, without knowing it, the government created arenas for the informal sector that emerged stronger than the formal economy, especially during the economic crisis of the 1980s.

The Arusha Declaration proclaimed universal primary education, primary health care, rural water supply, and villagization. In implementing these programs, the government expanded its labor force and tightened control over medium- and small-scale economic enterprises, including peasant agriculture, through political directives, watchdogs, and bureaucratic commandeering (Boesen and Koponen 1986). Because of the export of raw materials, especially cash crops and minerals, the economy was performing well enough to support this expansion and the nationalization of most social and economic sectors. The government nationalized the industrial sector as well as retailing, in which the government opened shops (*maduka ya ushirika*). At this point, the government was both the producer and the supplier of goods. The government decided what should be produced, the inputs, prices, and what was to be consumed.

The second five-year plan (1969–1974) was implemented at the beginning of Ujamaa. Seventy-five percent of the entire budget was directed to rural activities to strengthen villagization, only 15% was for agriculture, and 2% for the industrial sector (Havnevik 1993). Villagization policies insisted on self-sufficiency, focusing primarily on subsistence crops, and little on cash crops (Havnevik, 1993). Agriculture was the backbone of the economy. The cuts in budget allocation to agriculture and Ujamaa bureaucratic arrangements severely affected agricultural products (Scott 2008). Ujamaa policies ignored the potential of urban consumers to sustain agriculture. In the meantime, the industrial sector was struggling.

Manufacturing was supposed to connect with agriculture by processing crops and producing agricultural inputs. The industrial sector was also undermined by insufficient budget allocation. For that matter, the backbone of the economy, agriculture, was facing a double-edged sword. Even out of the tiny budget allocated for agriculture, a considerable chunk was allocated to crop parastatals, including marketing authorities and cooperatives (Boesen and Koponen 1986; Havnevik 1993). The third five-year plan (1974–1979) faced many challenges, including the rising cost of living, which had increased by 50%, and increasing urban-rural inequality.

The adoption of Ujamaa strictly discouraged the informal sector. Substantial restraints, both legal and moral, were put in place. *Uchuuzi* (street vending) was illegal, while the government established government-owned Ujamaa shops. The Ujamaa Village Act of 1975 was enacted to discourage exploitation in business. On top of that, the government conducted Operation Maduka (*maduka* translates to shops), wherein many unregistered shops in urban and rural areas were closed. Individual wealth accumulation was discouraged. Those who accumulated wealth through capitalism logic were regarded as *kupe* or *wayonyaji* (parasites or ticks) and as standing against the Ujamaa ideology (Havnevik 1993; Tripp 1990).

The Human Resources Deployment Act of April 1983 directed urban dwellers who were not legally employed and made their living through unauthorized petty business operations to have a labor identification card. Otherwise, they would be regarded as *wazururaji* (idles or loiterers) (Lugalla 1997; Stein 1985; Tripp 1990) and banned from cities in the amended penal code. On September 24, 1983, the Tanzanian workers' weekly newspaper, *Mfanyakazi*, published a list of activities identified as illegal and against the Ujamaa ideology, including the following: shoe shining; selling newspapers, buns, roasted cassava, and other foods in the streets; selling locally brewed beer outside of authorized premises; operating small vending stalls; car washing; special festival catering; and any petty trading that involved door-to-door selling (Kiondo 1990). The government's attitude toward the informal economy today, especially the use of force, is heavily influenced by this period.

ECONOMIC CRISIS AND THE COLLAPSE OF UJAMAA

The drought that hit Tanzania in 1973 and 1974 brought agricultural production almost to a halt. The global economic crisis of the 1970s, in conjunction with Ujamaa policies, led to a shortage of foreign exchange that further deepened the problems in the manufacturing sector. The lack of inputs for industrial production drove consumer product scarcity to its highest point. Basic goods such as food, cloth, sugar, soap, and cooking oil became scarce. By the 1980s, the government was also recovering from the 1978–1979 Tanzania-Ugandan war[2], and the Ujamaa policy and plans had hit the wall. Beginning in 1978, the GDP fell at about 3% per annum for three consecutive years, and inflation doubled to 30% and remained there until 1986 (Boesen and Koponen 1986; Mans 1994).

As production deteriorated and unemployment rose, people had no choice but to engage in informal activities outside the agricultural sector and formal employment. Given the economic situation, cooperatives could not pay peasants. Crops were bought on credit, which indirectly pushed peasants to sell their crops on black markets (Hyden 1988). Generally, production nearly ceased as the cost of production rose steeply, forcing the majority of people to seek additional sources of income. The situation perpetuated urban migration from rural communities as people set out to start small, informal businesses (Scott 2008). By 1977, food supplies were so low that they could only sustain 89% of the caloric requirements of the population. There was a scarcity of petroleum, and other imports were severely affected. Significant shortages of goods allowed for the rise of the black market (Edwards 2012; Gray 2013).

The informal repair culture and local craftsmanship we see today in Tanzania, which this book is about, are connected to this period. Local crafters and blacksmiths took over where manufacturing had failed, renovating available pieces of equipment and making household utensils and other vital commodities that the industrial sector could no longer supply to rural areas for production and home use. Local ingenuity became the new manufacturing sector, for example, by introducing a wooden beam plow or old rails for axles or making stoves from scrap materials to compensate for the lack of metal imports (Hyden 1988). The older crafters at Mahakama Ya Friji who were apprentices during the era remember properly how crafters were central in keeping life going.

The decline in production affected not only industries but also the household economy. Salary scales did not correlate with subsistence needs. The decline in agricultural and industrial production led to scarcity, which then led to the flourishing of the black market and price hikes for goods. By 1981, government interventions to restore production and rescue the economy had proven to be failures. There was pressure to accept liberalization as a condition for most aid and grants (Maliyamkono and Bagachwa 1990; Hyden 1980).

The burgeoning public sector that had grown sporadically in the 1960s came to a freeze, causing high unemployment; even for those with jobs, salaries stagnated for years. Many public employees, such as teachers, couldn't sustain their basic needs. While traders started operating in the black market, corruption became rampant in public services, accompanied by a brain drain among the well-educated. By the 1980s, the real wage was lower than it had been in 1963. In the span of five years (1979–1984), there was a 65% drop in the value of the real wage, while consumer product prices increased tenfold between 1976 and 1986 (Lugalla 1997).

A social survey conducted in Kariakoo, now Tanzania's leading business district, in the 1960s recorded unemployment at 11%; people engaged in petty trade and crafting were counted at 14%. Another survey in 1978 elucidated that a quarter of those aged 16–40 years engaged in petty trade. The economy worsened when the value of wages fell by 83% by 1988. The chronic shortage of consumer items, fuel, and food, poor infrastructure, breakdown of transportation, and lack of spare parts raised the cost of living, tightening individual and household economies (Brennan, Burton, and Lawi 2007). By the 1980s, informality had become a prominent source of livelihood.

From the late 1970s to the early 1980s, Mwalimu Nyerere refused to conform to International Monetary Fund conditions. He came up with strategies to help the economy recover, such as the National Economic Survival Program, which also went down the drain. As a result, the situation became impossible without external funding (Heinrich 1985). At this time, socialism was also disintegrating globally. The situation in the USSR and East Germany was alarming to poor socialist countries. Nyerere was quietly planning to retire and give way to changes. It was against his belief that Tanzania should liberalize its economy. He did not want this to happen under his government. In 1984, he announced "*nangàtuka*" (that he was leaving office) to allow another person to run the country (Shivji, Yahya-Othman, and Ng'wanza 2020).

LIBERALIZATION OF INFORMALITY

The collapse of Ujamaa, the economic crisis, and the Uganda war led to the inability of the Tanzanian government to pay for critically essential imports, fund social services, or pay salaries; hence, Tanzania reluctantly abandoned socialism and adopted liberalization. Nyerere left power, and the new president, Ali Hassan Mwinyi, was elected in 1985. In the same year's budget, the government officially removed agricultural subsidies. The government's involvement in the economy was supposed to be taken over by the private sector. New policies related to the Structural Adjustment Program (SAP) package were presented by the Bretton Woods institutions, and the government had to implement them to receive loans and donations from developed countries. Although the Tanzanian government and its people were against these policies, they had to be adopted to lift the country from its humanitarian crisis (Skarstein 2005).

With no more extended searching for economic saboteurs and racketeers, economic liberalization awakened and brought to light the informal sector. Tanzania had opened itself to foreign investment and adopted a free-market economy. The government withdrew from investing in the public and business sectors. The economic sabotage crackdown came to a natural death with liberalization policies. Mwinyi's policy was famously known as "*ruksa,*'" which comes from *ruhusa,* meaning "allowed or permitted." The ruksa policy was meant to open the country from strict Ujamaa to a free market. The country was opened, and business flourished with fewer restrictions from the state. This period was the zenith of the informal economy.

The era was also the beginning of consumerism in Tanzania. During this period, the boom in goods like secondhand electronics, clothes, shoes, and so on, which had been restricted under Ujamaa, flooded the country from abroad. At the same time, as a result of the economic downturn and SAPs, the government retrenched its public sector. Many public servants were retrenched to reduce government expenditures. Many of them invested their retrenchment benefits in small businesses (Mjema 1999).

At the same time, the government was no longer subsidizing farmers and cooperatives. The market was supposed to take over and rescue the agriculture sector. Nevertheless, the market could not save peasants from extreme poverty. Peasants could no longer afford farming inputs in the market. Crop yields dropped dramatically; hunger and famine became the order of the day. The exodus from rural to urban areas that had increased in the late 1980s rose steeply to new heights. At the same time, towns and cities were not prepared to host the burgeoning population, let alone manage their decimated budgets for services following the SAP. The high unemployment rate, poor social services, and degraded infrastructures could not keep up with population expansion. For these reasons, informality became a common way of earning a livelihood in urban centers (Bah et al. 2003; Degani 2017; Hill et al. 2014).

The fourth regime under President Benjamin Mkapa, who came to power in 1995, saw the informal sector, which had grown out of ruksa policies, becoming challenging to regulate, let alone collect taxes on. The laws to regulate it were inadequate. The President's office decided to engage in a long-term reform process by contracting the Institute for Liberty and Democracy (ILD), a Peruvian-based think tank chaired by economist Hernando de Soto. The ILD produced a draft plan called the Property and Business Formalization Program, known in Swahili as "*Mpango wa Kurasimisha Rasilimali na Biashara za Wanyonge*," or MKURABITA. The plan aimed at formalizing properties and business assets into a formal market economy (Steiler 2018; Steiler and Nyirenda 2021). Through MKURABITA, it was hoped that the informal sector would open new opportunities for loans and investments. However, the plan was not integrated with the National Strategy for Growth and Reduction of Poverty (NSGRP), which was adopted after Mkapa's regime. The two plans had principal contradictions in implementing microcredit schemes (Steiler 2018). Slowly, MKURABITA, which was not favored by the regime after Mkapa, faded in political discussions and met its natural death.

During Mkapa's era, the government also implemented decentralization by devolution, returning governance responsibilities to local governments. Among other things, collecting levies and organizing local markets were placed in the hands of local authorities. During this time, local authorities used *migambo* (parapolice forces) with no training to help collect levies from businesses in their localities. They used extra force in collecting levies, leading to innumerable fights between informal workers and migambo. Dar es Salaam was the epicenter; migambo *vs machinga* (parapolice vs street vendors) fights became an everyday experience for city dwellers (Brown 2015; Msoka 2005).

Soon after, the fourth presidential election, which put Jakaya Mrisho Kikwete in power, marked the start of a new era and another Pandora's box for the informal sector. At the beginning of President Kikwete's regime in 2005, urban authorities were instructed to remove informal workers from dense and busy areas. This act was purposeful in relation to the expansion of the city's infrastructure, including DART (Dar es Salaam Rapid Transport), which was a development target of the regime (Brown 2015).

During this time, municipalities set new policies for informal businesses to conform to municipal plans. The municipalities established designated places for informal markets. In Dar es Salaam, between 2008 and 2010, a five-story Machinga complex was built to accommodate approximately 5,000 businesses. The building had 4,206 stalls, eight storage rooms, 33 food vending spaces, and 68 kiosks. However, the location and business environment attracted neither vendors nor customers, affecting the businesses, most of which returned to the street. The building ended up empty, and the traders were back on the streets again.

TRANSIENT WORKSPACE

Mahakama Ya Friji is made of iron sheets. This reflects its precarious temporality, which is itself a reflection of these histories of relocating informal laborers. The materials of the building suggest that at any point, the workshop is ready to move to another location. This can happen in several ways; in fact, it has happened on several occasions: if the current arrangement with the property owner breaks down, if the neighbors complain to the city authorities, or if the city decides to relocate or close it. In addition, because of the nature of waste work, the workshop has to be strategically located in relation to other parts of the city that might provide scrap or buy crafted products. As a result, the workshop has moved many times before.

Ten years ago, before moving to the current location, the workshop had already moved between several places. When George and Mjomba Mjomba decided to start the workshop, they were working in someone's workshop, a typical arrangement for those who have relocated to big cities. George moved to Dar es Salaam from Tabora in the early 2000s. At first, he joined his uncle's workshop around Keko Magurumbasi, a neighborhood located between Kariakoo and the Keko industrial area. However, the income was not sufficient for his needs. He convinced Mjomba Mjomba, who was working at the same workshop, to start their own business. They opened their workshop in an open space between two houses. One of the houses belonged to George's maternal uncle on Togo Street in Kinondoni. They made a shade structure from UNICEF canvases, which were cheap and widely available after the Tanzanian government's 2009 closure of Burundi refugee camps. The workshop did not last long in this location after neighbors complained to the local government that it produced a lot of noise.

The workshop then moved to an abandoned hotel on Kisutu Street in Kariakoo. The hotel belonged to someone they knew from Tabora. They turned the cafeteria lounge into a working space. They were allowed to run their business for free in this building, paying only for security and cleaning services. However, the rapid expansion of Kariakoo forced them out. Eventually, the hotel was turned into a sausage factory. They had to move again, this time just a few streets away to an abandoned bus garage where a company had gone out of business. The garage had become unoccupied after the Dar es Salaam main bus terminal moved to Ubungo, 10 km outside of the city center. Once again, they were forced out of the garage due to the workshop noise. They then moved even further away to Msufini Kisiwani, 8 km south of the Dar es Salaam port.

Kisiwani neighborhood is in the south of the city, which is unlike the northeastern part that borders white sand beaches, five-star hotels, and affluent neighborhoods with homes for middle- and upper-class Tanzanians, who share gardens and walls with expatriates from embassies, international nongovernmental organizations (INGOs), and monetary institutions. Income differentiation and class become apparent while moving from these areas toward the southwest. Kisiwani is far away from the middle class, and materials are scarce and expensive. For this reason, the workshop could not survive in the south of the city. Proximity to the middle class, who regularly dispose of technology, is vital to the location of the workshop, as it means a higher supply and lower cost of materials (i.e., defunct devices).

At last, they rented a place in Kinondoni, just a few meters from where the workshop stands now. However, they lasted only five months in this location, as the property owner raised the rent erratically. They decided to move to an underdeveloped open space owned by the ruling party, Chama Cha Mapinduzi (CCM). They have entered into an agreement with CCM, that at any point when the party wants to develop the location, they can be asked to move. The iron sheet building reflects this readiness to move at any point. In front of the workshop, "*jiwe la msingi*" (a local party wing foundation stone) is erected, painted green with yellow writing, and has a flag post that holds a green flag on top (see figure 2.1). Several CCM election posters can be seen attached to the corrugated iron sheet walls. So far, 10 years have passed since they moved here.

This historical description of the workshop's movements shows that waste recycling sites like Mahakama Ya Friji do not just pop up. There are political and economic factors that determine their presence and their transience. They are not marginal but, in fact, places of deep connection to histories and politics, and social and material assemblages. Workers need to get along with property owners, create security networks, and be close to sources of materials to meet their customers. Otherwise, costs related to production, transportation, time, and capital become unbearable.

The presence of CCM support at Mahakama Ya Friji is also a response to the politics of the country. Discards labor, like all informal labor, has to be strategically positioned to the politics of the day. Waste itself is political and a tool for politics. Waste can influence political changes (Fredericks 2018). At the same time, as waste is a social classification, the politics of the day can determine what should be discarded and where. For the workers at Mahakama Ya Friji, supporting CCM is essential for their survival. Society has already classified their labor as a labor of waste, and so have those working with it. If they do not show support to those in power, they might become disposable in the face of local politics and local administration. Supporting opposition parties is not acceptable, regardless of whether some workers might be supporting opposition politics.

Figure 2.1 The Chama Cha Mapindunzi (CCM) Local Branch Stone at Mahakama Ya Friji. Photo provided by the author.

After President Magufuli's sudden death while in power on March 17th, 2021, his vice president, Samia Hassan Suluhu, assumed power immediately. President Samia took a U-turn on Magufuli's wajasiriamali identity cards, having a different view on the matter. The new regime came with new recommendations on how the informal sector should operate in Dar es Salaam and throughout the country. Since Magufuli's initiatives were not based on an act of parliament or other legal reform, they vanished after his demise. The identification card project had reached a dead end, too.

President Suluhu immediately directed regional commissioners to relocate street vendors to more convenient places to ensure fair competition with formal business ventures and to make sure they did not block pedestrian and vehicle flow during their undertakings. In Dar es Salaam, they were told to go back to the previously designated area, the Machinga complex building. However, because street vendors knew that the area did not attract customers, they continued to do business on the street until the military intervened and evicted everyone from the now-restricted areas.

This forced reallocation happened while I was in Dar es Salaam in November 2021. Most street vendors did not expect the profound force used to move them, including the military bulldozing business stalls and destroying goods. Despite the emphasis on relocation, wajasiriamali still takes risks by popping up on the streets when there are no police and running away once they show up. Even as the new president decided to bulldoze informal businesses in "not allowed areas," the workshop—with its agreement with the ruling party—was not affected.

CONCLUSION: INFORMAL ECONOMIES IN THE DIGITAL AGE

The liberalization of Tanzania's economy happened at a critical juncture, right when the country's economy was wrecked. The informal economy had already become the primary source of livelihood for most Tanzanians. As in much of the Global South(s), crafting and repairing with the few resources available became quotidian ways of being in the world. On the other hand, the information technology era had just started accelerating. The internet revolutionized digital communication, socialism eventually lost traction, and the market was penetrating every part of the world. Tanzania was not prepared for all these changes in terms of political infrastructure. The country had liberalized its economy while most policies were still socialist. President Ali Hassan Mwinyi opened everything to liberalization. There were no regulatory frameworks in place. To this day, Tanzanian authorities are catching up by patching policies and regulations to respond to liberalization's challenges, one of which is electronic discards.

In recent years, informality has taken a leap along with street vendors. *Bodaboda* (commercial motorcyclist) and *bajaj* (motorized tricycle) transport services have provided an informal avenue of employment for both educated and uneducated youth. The police have acknowledged that such enterprises have helped employ youth and decrease crime; however, it has become difficult to regulate (Heinrichs, Ehebrecht, and Lenz, 2017). These new informal ventures, accompanied by digitalization, have gained momentum and are changing the landscapes of the informal sector. The labor market, primarily digital-related jobs, provides skill-oriented opportunities rather than a myopic focus on education. In urban Tanzania, many young people are employed in digital jobs and informal digital businesses through platforms in delivery services, petty businessess through Instagram and Facebook, and the like. The transformation of the informal sector into the technosphere enables a new escape from government regulations.

The rise of the digital economy, which comes with information and communication technology devices, creates and cements certain types of informal labor, including repairing, crafting, and art. Workshops like the Mahakama Ya Friji that produce objects out of scrap, including charcoal stoves, BBQs, frying pans, and roofing materials, adopt newly available materials from electronics. Uhuru Street repairers are repairing our relationship with electronics and the planet. I see these gatherings of repair and crafting growing even more central in the handling of electronic discards.

NOTES

1. Ujasiriamali—the act of doing small business; Mjasiriamali—singular: A person doing small business; Wajasiriamali—plural: A group of individuals doing small businesses. Ujasiriamali, usually translated as entrepreneurship, is the Swahili word for a vibrant endeavor whereby people with little capital start small businesses. Instead of investors with significant capital or connections to capital markets, and instead of risk-taking, in Tanzania there are different perceptions and understandings of ujasiriamali. In general understanding, ujasiriamali refers to small business owners with little capital doing business. Several government officials, I spoke to defined ujasiriamali as conducting business and trade using creativity and innovation through selling and buying products or services, physical or virtual.

Important traits of mjasiriamali are that one must be motivated to change, in particular, motivated to change one's living conditions to something bigger. One must solve their personal problems to solve business problems. One must be ready to change at any time and take risks. A mjasiriamali is a creative and future-oriented person. They have to reach goals before others using unconventional methods. Typically, ujasiriamali refers to petty dealings operating in informal or semi-formal settings, including small and medium enterprises.

Ujasiriamali mainly aims for subsistence rather than further investment in a business, although some graduate to become large businesses. The wajasiriamali (informal sector actors) include innovators, hustlers, imitators, researchers, buyers, and so on. There are wajasiriamali with significant capital but operating in semiformal ways, whose businesses do not fully comply with formal business rules. What shapes or drives wajasiriamali to start their businesses is having a creative idea by looking at surrounding possibilities for trade. For that matter, ujasiriamali has no specific meaning; it depends on who defines it and for what purposes. According to legal and policy perspectives, mjasiriamali is defined according to the level of business compliance, capital, and operation model.

2. The Tanzania-Uganda war—1978–1979—started after Iddi Amin Dada, the then president of Uganda, invaded Tanzania. The war originated from the coup d'état in 1971, which removed Milton Obote, a close ally to the president of Tanzania, Julius Nyerere. Since then, Tanzania and Uganda were in constant diplomatic friction until 1978 when the war broke out. See more at Thomas, Charles. "Uganda–Tanzania War." *Oxford Research Encyclopedia of African History.* Jan 28, 2022.

Chapter 3

Repair Allegories

Phenomenology of Electronics Consumption and Repair

In April 2024, the European Union parliament adopted a directive on standard rules promoting electronic goods repair to tackle the increasing e-discards and accompanying problems. All member states should adopt and start implementing the directive within the 24 months from April 2024. The directive aims to address the following issues. Manufacturers are obligated to provide product repairs at a fair cost and within a reasonable time frame following the expiration of the legally guaranteed term. Consumers should be granted access to spare parts, tools, and repair information. Member states should promote consumer preference for repair by providing repair vouchers and funding. Online platforms ought to be established to aid consumers in locating nearby repair providers and businesses vending reconditioned merchandise. (Council of the European Union 2024).

The directive is seen as a positive step toward making customers find it simpler to seek repair rather than replacement. Repair services will improve in terms of accessibility, transparency, and appeal. In return, consumers will prioritize repairing over replacing goods, hence addressing more significant issues of electronic discards and greenhouse emissions, decoupling economic growth from resource use, and shifting to circular systems in production and consumption (Council of the European Union 2024).

Meanwhile, in Tanzania, repairing is an integral aspect of daily life for many. Unlike the EU, there are no specific policies or directives that incentivize consumers to repair their electronics or campaigns demanding the right to repair. Despite facing similar challenges, such as devices being designed for obsolescence, Tanzanian consumers always find a way to repair them. A fundi—a local artisan skilled in repairs—is always available to fix almost anything. These acts of repair or seeking out a repairer are not merely reactions to things breaking down. Instead, the repair is seen as an inherent feature of both an object's lifecycle and people's relationships with the object. Consequently, individuals often choose products based on the availability of repair services, the device's repairability, and the supporting infrastructure, even before a breakdown occurs. As stated in the repair directive, this perspective differs significantly from the EU region and offers an intriguing contrast in repair cultures.

Billboards advertising *fundi simu, fundi majiko, fundi cherehani, fundi bomba,* or *fundi ujenzi* are all over Dar es Salaam. Fundi means a person skilled in making, repairing, or maintaining things. Swahili does not have separate words for makers and fixers. Fundi is used as a pre-general noun followed by a specific term to indicate what kind of fundi. For example, fundi majiko means a cookstove crafter/repairer, fundi magari is a mechanic, fundi simu is a mobile phone repairer, fundi cherehani a tailor, fundi ujenzi a mason, fundi friji a refrigerator repairer, or fundi baiskeli a bicycle repairer.

Repairing is an everyday experience for many people in Tanzania. A fundi is always available to tinker with anything. However, acts of repair or finding a fundi do not only occur upon breakdown. Dynamics and possibilities of repair are considered and searched before acquiring an object. Tanzanian electronic consumers reflect on the object's repairability before purchasing and understand their specific assemblages. Purchasing power and poverty play an essential role here. Most Tanzanians cannot afford many new electronic products in the market. Chinese goods provide some new and affordable choices. Regardless of their means and choices, consumers need to know whether and where things can be repaired when they break down. The repairability of the object and a handy ready-to-repair, that is, the infrastructure for repairing, become central in consumers' choices.

This contrasts with a radical emphasis in other economies between the whole, intact, and functioning object, and the one that is broken, apart, and not working—in need of repair. In Tanzania, as I will show, for most people, objects are never whole and never really broken; even when brokenness happens, objects retain the potentialities of life, keep memories, and the intimate relations that the object gathers with the people. And there is a way to restore those relationships. There is a fundi always ready at hand to repair.

Ethnographies of repair have focused on issues pertaining to knowledge, borrowing from Levi-Strauss's *bricolage* (Harper 1987); material engagement (Orr 2016); repair in a structural sense, as in mending order (Henke 1999); repair as disruption in a material and political sense (Strebel, Bovet, and Sormani 2018); repair as intimacy and resistance (Gerasimova and Chuikina 2009); and sustainable world-building (Martínez 2017). In recent years, a growing cadre of researchers has also framed the labor of repair and maintenance in terms of care (Denis and Pontille 2014; Gregson, Metcalfe, and Crewe 2009; Graham and Thrift 2007; Houston et al. 2016; Martínez 2017), suggesting that practices of repair or maintenance (e.g., of infrastructures) can be understood as enactments of care. This "material care" (which also has implications for immaterial things, affects, and relationships) provides a salient point of departure for anthropological work on repair, materiality, value, and discards assemblages (Ntapanta and Medaas 2024).

The link between care and repair practices is vital to understanding the Tanzanian context. A narrative connection between the repair of electronics and care has arisen in societies where repair has been disembedded from consumption. Repair in those societies is not only about engaging material breakdown but is a form of resistance to consumerism, overuse of resources, exploitation of labor, political stance against techno-capitalism, and care for the environment. These forms of engaging with repair can be seen in repair movements around Europe that have been at the forefront of pushing for the EU Parliament to pass right-to-repair directive. Community repair cafés have also emerged in European cities like Copenhagen and Oslo, where people meet to repair their objects (see also Ntapanta and Medaas 2024). I specifically mention the EU here because, for most people in Tanzania, repair is an everyday response to everyday emergencies. There are no directives or policies from the state to incentivize repair. People repair because of the sociality of repair and its cultural and economic significance.

In Tanzania, repairing is a way of living; it illustrates how people are attached to things and the history and experiences they share with their possessions. Objects tell stories about their owners, their feelings, and enchantment. But most importantly, things must work for lives to continue. It is here where creativity and ingenuity arise. Acts of fixing and innovation demonstrate people's knowledge of making or fixing things and the need for life to continue in the face of failures.

In the face of everyday brokenness and infrastructure failure, repairing is an integral part of being in the world in places like Tanzania. The appropriation, improvisation, and contestation of the capitalist technosphere in Tanzania are integrated with histories, knowledge, and ways of being in the world and being with objects. Becoming waste is the very last condition an electronic equipment can reach in Tanzania. This last destination is avoided by all means. When a device is discarded, it is indeed beyond any potential. Devices that I encountered at the Mahakama Ya Friji workshop displayed traces of many acts and attempts of repair. Traces on devices reveal resistance to the techno-capitalist logic, where devices are made hard to repair with a shorter life span.

THE GREAT TECHNOSPHERE TRANSFORMATION

Why do practices of repair in Tanzania not slow down under the pressure of a neo-liberal market where planned obsolescence is a core feature? Is this simply a function of the relative lack of financial means of many Tanzanians? Recently, there has been a considerable increase in access to and ownership of electronics and appliances in Tanzania. Does this translate into the disappearance of repair work? It is interesting to note that current trends indicate otherwise. The streets of Dar es Salaam continue to be full of small repair kiosks. Three dynamics explain why repair has become even more prevalent. First, there is a peculiar way of being in the world with material objects. Repair practices are deeply rooted in people's ideas of "consuming," not in the sense of "to seize, destroy or take over completely" (Graeber 2011, 491), but rather the exchange between material objects and people. Because it is an exchange or relationship, people make sure that it continues, which necessitates care and repair.

Second, people's relationships, access, and scarcity of needed technologies reflect the histories of Tanzanians. As noted in chapter 2, Tanzania adopted a socialist model soon after its independence. In its doom years, the 1980s, Tanzania witnessed a serious economic depression. Industries died, and the country could neither make nor import many products. Tanzanians had no choice but to repair their possessions to keep life going (Kiondo 1990). Such historical experiences create and solidify ways of being in the world and here, very centrally, the repair of things. If families owned appliances or technological devices, they had to make them work as long as possible. They had to use and apply accumulated knowledge with their devices.

Third, the "great technosphere transformation" that has happened elsewhere by separating the histories and materialities of techno-objects from owners and society has not yet happened in Tanzania. This great transformation, which separated spheres of consumption and production (Polanyi 1944), did not take much root in Tanzanian society. Many Tanzanians still depend on the peasant economy for livelihood, whereby production is for subsistence. In wealthy consumer societies, the power of capitalism ushered in the demise of acts of repairing and altering relationships with brokenness. In the process, the repair was disconnected from consumption by capitalist design. The constant flows of new and constantly improving electronic technologies and products, along with aggressive marketing strategies, actively discourage repair and encourage people to continue upgrading to buy ever new and "better" gadgets and quickly discard the old and "outdated." Repair is no longer an option for consumers to consider. Brokenness is equated with complete dysfunction. Such a broken object is no longer part of wealthy lifeworlds and, therefore, quickly discarded.

Planned obsolescence is at the core of these developments. However, in Tanzania, people confront and circumvent planned obsolescence through repair. The transformation of the technosphere is contested in Tanzania by its design, spatiality, and relationality. Licensing and warranty do not interfere with Tanzanian consumers' relationships with their devices. Simultaneously, the Tanzanian government is failing to aid techno-capitalism. Most repair activities take place in the informal sector, where the government has already lost the battle to tax and regulate. Unlike in other economies, where governments serve as handmaidens to the electronic market through taxing repairs, the government interferes in repair matters and markets.

In Tanzania, objects do not quickly enter the sphere of the unfunctional, dead, or waste; rather, they travel back and forth between repair regimes, creating unique ways of being in and engaging with the world. Even though contemporary electronics are complex technological devices, as a result of the separation between people and the techno devices, through planned obsolescence, difficulty in repair, concealed and hard to open components, constant changes in the hardware and soft wares, and limited lifespan, repairers still manage to learn and figure out how each type can be repaired. Devices may take many trips to repairers, during which they are inspected, opened, soldered, and screwed. A device might be passed between individuals and repairers until no expert can salvage it.

EXPECTATIONS OF REPAIR AND REPAIRABILITY

I vividly remember the evening of July 16, 1994. Tanzania had just adopted the Structural Adjustment Package, a new market economy was emerging, and the country was to hold its first multi-party election a year later. After dinner, served at 8 p.m. sharp every day, I was sitting with my father, listening to the news on our Panasonic radio. The FIFA World Cup was heading into the finals. Brazil was going to play Italy early in the morning of the 17th. This was the first World Cup that took place in the United States of America, and Nigeria played for the first time in the tournament. The tournament had other historical firsts: Unified Germany played as one team for the first time, and there was a new Russian team following the collapse of the Soviet Union. Many Tanzanians were able, for the first time, to watch the World Cup on television after President Mwinyi's *ruksa* (liberalization) policies had marked the end of Ujamaa (socialist period).

My father had made sure the radio had new batteries to avoid missing any action. Several hours before kick-off, my father warned me that I wouldn't be awake. He asked who I predicted would win the World Cup. To this day, I still wonder about my answer. I had many heroes on the Brazilian team, but I rallied for Italy. After an inconclusive match, the winner had to be decided by a penalty shoot-out. For the first time ever, a team won the World Cup on penalties. When I woke up the next day, my father delivered news of my defeat. Italy lost the World Cup, and Roberto Baggio missed the last penalty.

But my morning was made when my father announced that for the next World Cup to take place in France, we were going to watch matches on TV in our sitting room.

Unfortunately, that didn't happen until 14 years later, in 2008. I was in the third year of my bachelor's degree when my father had raised enough funds to buy a television. He traveled by bus from Tabora, our hometown, to Dar es Salaam to buy a TV set, a satellite dish, and a power generator. We needed a generator because there was no electricity where I grew up. As the eldest son studying at university, I was supposed to accompany him to Kariakoo for shopping. Before going to Kariakoo, I asked him what he envisaged buying. He wanted a used Hitachi TV, a new generator, and a satellite dish. At the secondhand shops we visited, there were other TV brands, including Panasonic and Sharp. Some were cheaper than the Hitachi he wanted. However, he insisted on buying the Hitachi.

My father preferred the Hitachi TV because there were repairers in our area who could fix this brand. He was afraid of buying other brands because it would be difficult to find a repairer when the TV broke. People think of breakdowns when purchasing devices, and repairability plays an important role in decisions about a concrete product. From the socialist policies that limited access to electronic devices, to economic hardships that preceded liberalization, to the emergence of counterfeit products in the market economy, people search for secondhand products from Europe and the Americas. The wear and scratches on devices assure people that the product is usable and durable. However, this comes with the expectation of breaking down.

This way of thinking, with the object's promise to break and its repairability, is very different from what I experienced in Denmark or Norway. While I was writing my PhD dissertation, I discovered five defunct coffee machines in my garage that I had used and discarded between 2018 and 2022 (see figure 3.1). On impulse, I decided to post about them on Twitter, wondering if writing about electronic discards leads to even more production. Comments from my network, especially in Tanzania, wondered if the coffee machines could be repaired or if I even cared to repair them. If there were repairers, how much would it cost if I tried to fix them or get spare parts from one machine to fix another one, or could I export them to Tanzania, where they could be fixed? The responses from my networks in Europe showed disappointment in consumerism and asked if I cared for the environment. They also blamed techno-capitalism. Concerns were directed toward regulating planned obsolescence, standardizing quality and designs so that people can fix things themselves, incentivizing repairing through tax breaks, the importance of creating an environment conducive to a circular economy, and the overall lack of care toward objects and the environment at large.

Figure 3.1 Defunct Coffee Machine in My Garden in Denmark. Photo provided by the author.

If my father had bought the coffee machines, he would have sought advice from local repairers about which brand is economical and can be repaired when parts break, who could repair it, how long, and at what cost. When I bought the coffee machines, I thought only about the price, regardless of whether they could be repaired. And if they broke, I sought the next cheaper option to buy because I knew there was no one around to repair them.

When shopping for electronics, buyers in Tanzania are concerned with brands. They care not only about the symbolic status presented by brands but, very importantly, how likely it is that a device will break down in the future and how easy and costly (or not) it will be to repair, just like my father's search for the TV. In 2022, when visiting my parents in Tabora, I brought a used iPad from Europe for my mother to simplify our communication. She was very happy but also worried about how she could find a repairer in Tabora and how much a repair would cost. Her concerns were about this specific, not well-known device among the local repairers. As opposed to Martin Heidegger's phenomenological allegory of the broken hammer, in which everyday things become invisible to people when they function properly, Tanzanians think about the object in terms of when it might fail to perform and do not wait until it fails (Puthussery 2019; Sormani, Bovet, and Strebel 2019; Strebel, Bovet, and Sormani 2018).

Electronic devices have multiple ways of being in the world. Their encounter within networks evokes different relations that are unique and relative to those assemblages. In places where the great technosphere transformation has occurred, things might quickly end as discards when the first breakdown happens, which means the end of the object's life. Elsewhere, the device might get repaired or start a new life stage through reuse or repurposing.

The lifeworld of an object is not a generalized occurrence but includes a multiplicity of quotidian experiences in temporalities and spatialities. Being in "this world" is an arena of constant improvisation of actions upon brokenness or breaking that necessitates responses to what the world is or might become (Simone 2022). Analyzing "this world" from a Deleuzian perspective, AbdouMaliq Simone argues, "While that which is broken might seem to be the very basis for a hoped-for resolution in a new world, Deleuze sees this brokenness, of being without a world, without possibilities of a progressive, sustained relationship among things" (Simone 2022, 26). Tanzanians share in a desire for a new world that settles in long, unbroken durée. However, they are aware that the new world is a far reach. They are assured of this world—that they must prepare and engage with its promises to break. Repairers are ready in every corner of urban Tanzania to repair brokenness so that this world continues to exist.

Things share space with people's experiences of being in place and time, being in this world. Time here means the present and the continuum from the past to the future—people's engagement with this world in the past shapes their present and future encounters with the world. Things have agency themselves in shaping the experience of the world. People influence the world through their actions and the logic of shared experiences of this world through things they engage with. Heidegger proposes that in everyday engagement with things, they tend to disappear and become invisible to people (Puthussery 2019). When they function correctly, the paradox of everyday engagement with things tends to settle in the background until its function is disrupted. Only then does the agency of the thing or its being come to light.

However, this Heideggerian paradox loses its depth once brought into the light of consumption, repair, and discarding of electronics by Tanzanians. Like my father, the majority of people in Tanzania think of the failure of the object even before acquiring it. Before that moment, they must know who can fix it, how long it will take if spare parts are available and affordable, and what the cost will be. Most Tanzanians consider all these factors before acquiring electronics. Electronic consumers in the EU are also well aware of planned obsolescence embedded in the devices they buy or own. They know the devices' life spans are short, expensive, and painstaking to repair, and it is cheaper to buy a new one. During my coffee machine purchases in Denmark, it did not occur to me that I would need to repair it. Though I knew they were going to break each time, I bought one. That is the central reason why my friends in Europe asked if I cared for the environment. In this very particular world, the hope for this world that is sustainable has become obsolete, the new world has arisen, the continuum from the past to the now has been systematically curtailed, and repairing this world has just become a mere desire. The great divide between technology and society has already happened.

The Freedom Street: Ethics and Terrains of Repair

One of the busiest streets in Kariakoo, Uhuru Street, which translates as Freedom Street, could be considered Tanzania's tech hub. Uhuru Street is mostly lined with electronic stores. Due to these stores, the street has also drawn mobile spare parts retailers and software specialists, many of whom are recent IT graduates without jobs, as well as repairers who are typically do-it-yourselfers without any formal training in electronics.

Retailers of spare parts import parts from Asia, primarily China. They also purchase obsolete electronics to dismantle and resell the individual parts. Additionally, Uhuru Street draws large crowds of shoppers who cram past one another through its hallways to purchase various devices based on their income. There are hundreds of repairers in the corridors between the shops, most of them mobile phone repairers. The halls are lined with small tables, and the pedestrian lanes are shaded by multicolored umbrellas that protect the kiosks from the scorching heat of Dar es Salaam. The colors make the entire street look gorgeous. Tables must be removed at night and returned the next day, resulting in a daily change of only a few yards. This is the street where my repair Walimu, Hussein, and Msomali, are located. Every time I visited them, I had to walk for a few minutes before finding where their tables were (see figure 3.2).

In Tanzania, repairers are navigating terrains of formality and informality, precariousness in their relations with the state, spontaneity, and plasticity when it comes to repairing spaces and methods. Within such kind of landscape, certain ethics emerge, solidify, and apprehend. They are not trained in schools, do not have licenses, nor do they have manuals to look at. They innovatively find each problem and fix it by any means they can. Sometimes they cannot fix an item. In those cases, they propose to the customer to find someone else. However, when they believe that a job is difficult but that the device can be repaired, they kindly ask the customer to leave their device and to come back after several days. They will also make sure the customer knows what is at stake.

At the same time, repairers must also learn about new devices and new problems. In these cases, they will warn customers that the device may not be fixable because they have not yet learned yet to repair that type. Then

Figure 3.2 Uhuru Street Kariakoo Dar es Salaam. Photo provided by the author.

later grapple with the problem through trial and error. Many times, they will manage to fix it, but sometimes they fail. This is also how they learn about new devices and new problems. There are no other ways of learning than by doing. Once one of them learns how to fix a particular device or problem, others will bring that work to him. They will sit, observe, and learn so that the next time the same problem comes to them, they can fix it.

Repairers might seem slow or sluggish during fixing. They will keep chatting while the customer is sitting on the bench, waiting. If a new customer comes by, they will put the previous device aside and talk to the new customer to understand the problem. They might ask the customer to come back later if the problem is hard to fix or ask them to sit on the bench and wait if not. They operate on their own logic which might be difficult to see at first glance. Here, repair work is tied to classical ways of doing repair. The aim and labor are entwined with fixing the device and are not time used to repair (Harper 1987; van Maanen 1990). The sluggishness might get misinterpreted, probably because of the marginal gains of their labor; rather, the slowness is, on one hand caring for their customers and the devices. A way of giving their ears to listen and identify the device's problems. And on the other hand, chit-chat, in many cases, involves problem-solving and better ways to fix devices.

The more the repairer makes mistakes or fails to fix them, the more one loses customers and trust. As Hussein told me, the most important elements in repairing, at least on Uhuru Street, are the moral values of honesty and trust. "When a customer comes here, we tell them the truth; if I can't fix, I tell. If I can try or if I can fix it, I also tell." He continued, "Customers who travel all the way to Uhuru Street, they trust us, they know we can fix their phones, we have to listen so that we do not do mistakes." Depending on the repairer, the popularity, and the area of operation, it may take longer for the customers to get their devices fixed. Honesty, listening, trust, and care are essential moral values of the repairing trade (van Maanen 1990). Even though there is a kiswahili saying, *hakuna fundi mkweli bali kinyozi*, (mafundi of all types, trust only the barber). Nevertheless, electronic repairers have high ethics in telling what they can and cannot do.

This kind of sociality found at Uhuru Street cannot be found when one goes to an authorized dealer. Trust is directed toward the dealer's shop, not the person who fixes the item. This trust is built on manuals and training. Repair is directed with steps on what to do and how the device should be fixed (Houston et al. 2016). Usually, sociality between the customer and the fixer is restricted. And when an error occurs in the authorized dealership, the customer is often compensated, which also happens sometimes at Uhuru Street. There is no imagination of future relationships between the customer and the repairer. This contrasts with Uhuru Street, where repair happens through collaboration and improvisation, and is also about learning. On many occasions, customers let it go when repair fails, and they might return in the future. Sociality and social trust are more central to the encounter than even the technical ability of the fixer.

Repair in Dar es Salaam is a societal project. Things have to be fixed. To fix a device is to fix many other societal issues that are connected to the service of the device. And, of course, on the bigger picture, fixing our planet. From my observations in the EU, I came to understand repair is enacted as a form of care in a deeper sense. Repair has been extensively integrated into the narrative of assuming responsibility to alleviate the negative effects of consumerism and environmental deterioration. In the EU, relationships with the objects that necessitate repairing have been disembedded. If we desire to do repairs, there must exist a very profound motive that goes beyond only addressing the brokenness. There are financial means, comparatively, to purchase a more recent and less expensive item as a replacement for the damaged one. This narrative link highlights the profound connection between repair and responsibility, which might stem from a sense of complicity or serve as a means of proactive engagement in the face of seemingly monumental challenges, encapsulating the spirit of "acting locally, thinking globally" (Houston et al. 2016; Ntapanta and Medaas 2024).

Claims to planetary care and the morally charged political vanguardism of repair that we know from some communities in the EU do not entirely elucidate the meaning of my observations in Tanzania. In this chapter, I have carefully observed and tried to understand the processes in which repair strategies develop, are conceptualized, and implemented, and how resistance to the capitalist technosphere manifests in Tanzania. I have keenly focused on knowledge, practices, and imagination from Tanzania that provides an avenue to think of repairing and possible ways to circularity.

CONCLUSION

My father's Hitachi is still in our house and working even today, though it is no longer in the sitting room. Since then, he has bought another updated flat screen television set, but he has managed not to let the old Hitachi go. Part of me still remembers the trip to Kariakoo. The TV, although it has lost its color, still works well. It shares an attachment and history with my father, me, and probably the whole family. Even though it is not often used, simply by being there, it tells a story about the family and the past and reminds us of the inspiration for the future. Part of its continued existence, that my father cannot let go, is the expectation of repair. One day it might come to use or provide spare parts for repair. Even though it is an old model, an outdated technology, that will not provide spare parts for a current smart TV, that expectation stays.

With the many soldering and screwing marks that the Hitachi TV has, it tells a story of repair. The trial and error, successes and failures of bricoleurs who did not attend technical schools to learn the functions of electronics. Moreover, it reflects the incremental knowledge that repairers have and must acquire to fix the device, but also to fix this world. The ways Tanzanian electronic consumers think of repairing even before making their purchase are specific and must be understood in their particular contexts. Of course, there are issues of purchasing power. Most Tanzanians cannot afford some of the new electronics on the market. However, with supply from China, there are many new and affordable brands. Nevertheless, even if the device is new, people need to know where it can be repaired when the device breaks down. Purchasing power is not the substantial factor here, but rather having a handy ready-to-repair.

Brands known for being expensive and difficult to repair, like the iPad I gave my mother, are not preferred. My father searched for a Hitachi TV because it was easy, cheap, and many repairers could fix them. Brands, especially those difficult to repair, like Apple devices or Europe-manufactured cars, project a special social status. When one owns Apple products or a European car, they are considered upper class because of what the brand projects and because they are expensive to repair. The owner is expected to have an excellent income to own such brands. Few repairers know how to repair these brands, and spare parts are scarce, making repair expensive.

In Tanzania, repairing emerges neither as a political movement nor out of concern for overconsumption or waste, nor hopes of the new world that has yet to come. Repair emerges as a practical matter with respect to a specific object, working against disruptions that affect the functionality of the device and lifeworlds in this world. It is embedded in consumption. I see repair practices in Tanzania as countering alienation, decolonizing, and affirming self-reliance in certain ways. Clapperton Mavhunga (2017) argues that African history is laden with examples of adjusted crafts and traditions that help people solve their daily problems. Instead of seeing electronics as an innovation coming from outside of Africa, tinkering and reoccupying these technologies to fit people's needs is an innovation from within Africa.

Even though, as mentioned above, some brands are not preferred because of difficulties in repairing them, repairers in Dar es Salaam will still find ways to make devices function. Repairers will even travel to scrapyards searching for spare parts. Devices make several, if not uncountable, trips to repairers before they are no longer repairable. Then, they are stored somewhere in the house until scrap collectors come to pick them up in one way or another. The next chapter follows these objects through their next stage: as scrap for collectors, dealers, and recyclers.

Chapter 4

Gatherings

Coming Together of Discards Labor

Mahakama Ya Friji is a workplace where labor is exerted; "electronic discards are reshaped, one life ends, and a new life starts," as George once said. Mahakama Ya Friji is a transient space; materials are ingeniously revalued, repurposed, and reshaped. At the workshop, defunct objects acquire new values and functions. Lives are assembled here. When I say lives, I mean life in its totality: humans and the nonhuman, that is, animals, bacteria, chemical compounds, metals, and so on. At the workshop, lives are constantly shaped by this gathering—the coming together of different lives.

Mahakama Ya Friji is not a cast-off place; it is not a "Sodom," as unregulated electronic discards recycling places in African cities are portrayed by the international media (Adjei 2014; Weigensamer and Krönes 2018). Recent literature on discards spaces reveals that discards labor is an outcome of colonial and postcolonial arrangements (Nguyen 2019), ethnicity, and class divisions (Gregson et al. 2010; Millar 2018; Millar 2008; Millar 2014), racism and caste systems (Gregory 1997), politics (Fredericks 2018), innovations and technological changes (Gabrys 2011), and global interconnectedness (Lepawsky 2018; Lepawsky and Mather 2011). While the literature on discards emphasizes global relations, history, and power, it still tends to focus on harm and marginality. Of course, some more than others. The story of Mahakama Ya Friji is different. What do I mean here? Spaces like Mahakama Ya Friji are seen as the margin of techno-capitalism. Places where the tides of afterlives, to which the people who live and work there are surplus. They are also seen as places of pollution, both in terms of toxicity, wastefulness, but also as spaces with moral depravity (Douglas 2003). The workshop is not a collection of poor black bodies. It is not a mere recipient of technology, a site of backwardness in techno-capitalism, or a space out of place, or a wasteland. The usual narrative is that these places are separated from their cities, found on empty grounds and apocalyptic sites. The workshop is also not, or not only, a place of toxic violence. One could view the workshop as a "disposable" place, a place for disposable materials and bodies. However, I attend to the workshop in a postcolonial city as a workplace.

The workshop is deeply interconnected with the city through the production of artifacts that are central to the everyday life of Dar es Salaam dwellers. Mahakama Ya Friji and other scrap workshops produce cooking pans, roofing materials, and charcoal stoves. Ninety percent of all households in Dar es Salaam use and possess a charcoal stove that is used for cooking. Even those 10% who use other sources of energy, wherever they dine out, definitely have some of their meals cooked with charcoal stoves. Charcoal stoves, on the other hand, are not produced on an industrial scale. There is no factory for charcoal stoves. All charcoal stoves are produced in small and medium workshops like Mahakama Ya Friji.

The workshop is also connected to the country's vibrant metal business and construction sector. Metals like copper that are extracted from electronic discards are exported abroad, hence Tanzania receives forex out of the business. Other metals like iron are used to produce metal rods central to the real estate and infrastructure construction boom Tanzania is experiencing. For these reasons, the workshop is also connected to transnational technocapital—not on the margins but, in fact, at the heart of these assemblages.

However, as I have mentioned earlier, the workshops are framed as informal spaces that the government wants to formalize or does not want their existence, zones of electronic discards, pyropolitics, and pollution. They occupy a precarious position in the city in terms of their geographical location, denied rights to work in the city, and continuously facing expulsion, stigma, and violence.

This chapter explores the gathering as a coming together of social and labor relations within the workshop. These gatherings do not just happen. As a workplace, the workshop has its bureaucracy, very close to the Weberian model, and politics. Hierarchy and power dynamics are reflected in workstation divisions, possession of vital tools, social capital, and income.

GATHERING LABOR: THE COMING TOGETHER OF INTERDEPENDENT NETWORKS

The hierarchy of the workshop is multi-layered. There are around 20 scrap collectors/gatherers—almost all young, in their late teenage years to their 20s—as well as 25 to 30 charcoal stove crafters and five scrap dealers. Most of the workers are first- and second-generation migrants from rural areas. Each group, from scrap gatherers to scrap dealers to apprentices to master crafters, is interdependent. The gatherers bring in the scrap materials needed by the crafters and dealers. In addition, both scrap dealers and collectors need the crafters because of the crafters' agreement with the ruling political party—Chama Cha Mapinduzi, the landowner.

To understand labor relations, processes, and specializations in the workshop, ethnic attachments and kinship relationships play a vital role. Migration from rural areas operates through networks (Parry 2003). Usually, migrants move from their village to the next town, the district centers in the case of Tanzania, then to regional centers, and finally to the cities, Dar es Salaam being a prime destination. The pioneers migrate first and then invite their kin and neighbors. For example, most crafters come from Tabora, an inland region that is one of the poorest in Tanzania; George and Mjomba Mjomba came first, and their brothers and friends followed later. This is the reason that specialization in the workshop is grouped by ethnic relations, creating an urban politico-economic organization (Cohen 2014).

The workshop's organization, morality, values, and hierarchical arrangements are embedded in these assemblages. The master crafters, including George, Mjomba Mjomba, Mwanawane, and Hassan, come from the same *kabila* (a group of people sharing the same language and culture) called Nyamwezi. The domination of the Nyamwezi at the workshop does not mean that they are waste workers. Instead, kinship relations organize the division of work at the workshop. Among the discards gatherers, there were no Nyamwezi. In contrast, even the youngest craft apprentices are from the Nyamwezi. Among the crafters and their apprentices, only two came from other ethnic groups. The same applies to the scrap dealers, as two out of the three came from the Chaga ethnic group, from the area around Mt. Kilimanjaro.

Workers' spaces within and around the workshop reflect both hierarchy and interdependence, as well as the kinship relations shared by many at Mahakama Ya Friji. George, Mwanawane, and Mjomba Mjomba, who form the workshop's leadership caucus, have their stations at the front of the workshop, under the neem tree where the electricity meter is placed. Some of their jobs, apart from running the workshop, are to monitor the use of electricity, collect money when their power meter dries, and contact people about workshop affairs. These master crafters are considered the skilled personnel at the workshop. On top of holding the lease agreement with the landowner, they occupy the workshop hierarchy's top level due to their skill. Although the scrap dealers form the workshop's tycoon class, their business is still dependent on the crafters. If the crafters move, the dealers will have to move with them.

The workshop floor is divided into three parts. The front part lies under the shade of a neem tree, occupying a strategic position for the master crafters to meet customers and offer an arrival port for scrap collectors. Their occupation of this strategic position in the workshop reveals their position in its power strata. Ramadhani, another master crafter, has a station in the middle of the workshop where he works together with his two apprentices. Ramadhani is the newest master crafter. Though he could get a spot in the front, he prefers working inside the workshop. In the afternoons, when scrap collectors return from their trips, crafters gather in the front to salvage materials that have been collected. Those in the front get to inspect the scraps before those inside. They are also the first to meet stove customers, negotiate prices, take orders, and estimate products' delivery time.

Kinship and hierarchy intersect in various ways. The younger Samwel, who had just finished primary school and was without any prospects for continued studies, had joined his grandfather, his uncle Mwanawane, and his father Charles to become the third generation of crafters in his family. Samwel and his grandfather occupy the southern corner inside the workshop near the television. Charles and Mwanawane make deals with customers, sketch plans for stoves depending on customers' needs as described, and then send these to Samwel and his grandfather to make the parts.

The further back you go inside the workshop, the lower the workers' positionality on the hierarchy. Hassan, an early member of the workshop, had returned after a two-year employment stint outside the workshop and has a station next to Samwel and his grandfather. I was told that a conflict had led Hassan to leave the workshop, which came from an accusation Hassan had

Figure 4.1 Dismantling of Air-conditioner on the Front of Mkali's Warehouse. Photo provided by author.

made against George. George was the workshop chair at the time, and Hassan accused George of treating his apprentices unfairly, leading to Hassan's departure. He found a job as a technician at a company that makes aluminum frames for windows and doors.

Hassan told me that the lack of flexibility that he used to have at the workshop, the inability to take on extra jobs or take breaks when he wanted, and a lack of companionship forced his return. Hassan's return to the workshop is vital in understanding labor conditions there. The case serves as a backdrop for the idea that unregulated recycling jobs are undesirable and the last resort for people. Hassan left formal employment because of the benefits of working at Mahakama Ya Friji. Not only in the sense of income, but other values, such as friendship, power over one's labor, and control of the production process. However, when he returned, he had lost his former position at the front of the workshop. He now had to occupy the inside and recruit new apprentices. He found four new apprentices, who work behind the workshop under a shaded workstation along with George's apprentices.

The three scrap dealers' warehouses are also located in front of the workshop. Mkali and Anton's warehouse is on the right side of the workshop. On the left side of the front of the workshop, Babu Ali's warehouse is located close to the entrance. During my first fieldwork visit, Babu Ali came only a few times a week because he had other business. However, he was at the workshop every day during my second, during the COVID-19 epidemic, and third visits at the end of 2021 and 2023. I never learned what other business he was doing, but I was told at the latter moment that the other venture was not doing well because of the effects of the epidemic, which was why he was at the workshop more frequently.

At its far end, the workshop borders the Dada grocery and a warehouse for two dealers who are a couple. I had few interactions with them, which means I did not get to know much about their business. The warehouses are in front of the workshop because there is enough space by the road to accommodate collectors and their trolleys, and weight machines and because it provides easy access for trucks to load.

Scrap gatherers are somehow at the bottom of the hierarchy because they have no formal place in the workshop. Most of their labor takes place beyond its walls, only in the morning when they pick up the trolleys and start-up money, and, in the evening, when they return (see chapter 5). However, the gatherers are central to all activities at the workshop. They are the material and value gatherers. Without them, crafters will not get the material needed for their craft, and scrap dealers will not do business. More centrally, they are the valuators. They are the first to encounter devices from households, and they must evaluate accordingly so that the crafters can afford the material, and the dealer can get the metals.

Gatherers have vast networks of suppliers. The same applies to the scrap dealers; each gatherer has their own network and is connected to a dealer. At the same time, repairers and crafters are connected back to the dealers. Devices are gathered at low cost by the gatherers, who know the approximate actual value and how to negotiate better prices. When objects arrive at the workshop, crafters salvage what they need, repairers pluck functional materials, and the rest stays with the dealer. This interdependency diffuses power among the groups (gatherers, dealers, and crafters), as each group's labor depends on the others. At the same time, there is still a hierarchy embedded in this dependency.

The minimal hierarchy, ways of organizing labor, and interdependency as a form of highly specialized and skilled labor. As I have described above, the labor organization is a contrast to the typical understanding of "marginal" informal discards work. There is more to the value of a labor organization that is deeply rooted in collegiality, reciprocity, and interdependency between workers that spark networks of creativity, ingenuity, and long-term friendship. One final worker at Mahakama Ya Friji deserves mention, one who inhabits the space differently by day and night. Standing outside human hierarchies but still providing valuable labor to the workshop is Dog Master.

MORE-THAN-HUMAN DISCARDS LABOR

At Mahakama Ya Friji, there is another member, a dog, with human workers. During the day, she usually sleeps under piles of scrap deposited against the side of Mkali's warehouse. She sleeps most mornings and afternoons. In the evenings, she starts roaming around the workshop. She rarely crosses beyond the perimeter of the workshop. I assume that she feels safe around people she knows—her kin, members and colleagues—and the minimal traffic of cars and motorbikes that pass by the front of the workshop now and then. She is called "Dog Master" and is a loved and feared member of the workshop. Like the chipped dogs in Chen Qiufan's (2019) novel *Waste Tide*, Dog Master is also a security guard who acts against intruders.

However, she is also a good companion at the workshop. They adopted her when the workshop moved to this location. Everybody cares for her in their own way, like giving her part of their meals or water. They get annoyed if anyone tries to abuse her. I also developed a close relationship with Dog Master. She knew my arrival time, our lunchtime (as I often shared with her), and developed an urge for my petting. In return, when I was tired and took a break or a nap on my seat, she would come and lie next to me.

I am told that one night, a thief broke into the workshop while the guard was sleeping. Dog Master grabbed and held the thief for more than 30 minutes until the guard appeared to apprehend him. Once the sun sets and night arrives, Dog Master is no longer the dog that sleeps during the day. She becomes a guard and a hero for everyone at the workshop. She guards the workshop well. They have more trust in her than even in the guard. At night, she will bark loudly even at members of the workshop.

Discards labor is more-than-human labor. Humans' animal companions participate in discards labor as infrastructure for transportation and facilitating metabolism, or by providing care and companionship to the humans working with discards (Fahmi and Sutton 2010; Doherty 2019), and by protecting their workshops. Animals are discards workers at the downstream of capitalism. They are, in many cases at the center of value production (Barua 2018). What goes on in discards labor also includes the invaluable work of other species (Haraway 2012), which depend on each other during discards labor. As they become discards kin with human discards labor, they fall into the same stigmatized category as their human fellows. They also experience the brutal end of life downstream.

When I visited the workshop at the end of 2021, Dog Master had given birth to a male dog. He was two years old when I arrived. He would have been conceived around the time I was doing my first fieldwork. The male dog, who was named Simba, joined the workshop as a second-generation worker like his mother. He is a loving dog during the day and a strong guard at night. At night, Dog Master guards the back while the boy dog is in the front of the workshop. The whole street knows the dogs. One Mama Lishe told me, "During the day they sleep like useless dogs, try to come late in the night. They are different creature." When I returned to the workshop in 2022, Dog Master had passed away. She left three sons whom she had trained well to become the new generation of Mahakama Ya Friji guards.

In the same way, the discards labor of other species is unrecognized. Tools used in unregulated electronic discards recycling are perceived as inferior and a source of pollution by environmentalists. At the same time, these regarded primitive tools are central to salvage accumulation by the capitalist system.

POLITICS OF TOOLS

Discussions about proper recycling of electronics are typically based on the methods used. However, in later chapters (5 and 7), I argue that it is not about the methods; the tools used define the methods—the more "advanced" the technology, the more formalized and appropriate the methods will be, and vice versa. However, when technology advances, tools tend to concentrate in fewer hands, monopolized by those with capital. This condition is also occurring at the workshop. The monopolization of the means of labor is the beginning of the alienation of labor from production processes and products and of the exploitation of labor. The use and ownership of tools define labor relations and labor compensation (Wendling 2009).

The sociological taxonomy of workers at Mahakama Ya Friji is premised on the tools and techniques used to recycle discards. Tools imply control over the production process: control of and over the time and space of labor; control of choice on how, when, and what to do. Tools are central to labor assemblages and relations with material objects. The relationships on the industrial shop floor are parasitic between the owners of the tools i.e., machines and the workers. It is distinctive at Mahakama Ya Friji. Tools are the junctures, the gatekeepers of the energy and skills of the worker who interacts with materials to produce, transform, or transport a craft, an object, or a product, the same agency as on the factory shop floor. However, tools do not give the workers on the factory floor the same power and freedom as they do at the workshop (Wendling 2009).

Tools are emancipatory. Owning vital tools needed for production is owning one's labor. At the Mahakama Ya Friji, the crafters and the scrap dealers own primarily necessary tools. They impact the flow of labor at the workshop. At the workshop, tools can influence, by gatekeeping, who can produce at certain moments. What defines a crafter's, dealer's, or a gatherer's labor are the skills they must acquire and the vital tools necessary to implement those skills (see also chapter 5 and 7). Even though workers at Mahakama Ya Friji might enjoy a certain degree of freedom and control over their labor and the means of production, they are still entangled in the structures and networks of capitalism. These interdependencies are grounded in the capital (i.e., the tools, financial capital, and the start-up funds).

Scrap dealers and scrap gatherers work interdependently in several ways. There are no junk places, and scrap does not reach the landfill. Most scraps are gathered from households and businesses. This can happen in two ways: first, by buying from owners, which can be households or businesses, or by getting scrap for free, especially from wealthy neighborhoods. There is no assurance of getting scrap for free. Therefore, gatherers need to have money to buy scrap in case they do not find it for free. For this reason, a scrap dealer to whom a gatherer is attached must give the gatherer a daily start-up loan. A *kianzio* (daily start-up loan) is money that a scrap dealer gives to a gatherer, which the gatherer then uses to buy scraps. Upon returning to the workshop, the gatherer will sell whatever they have collected to repairers, crafters, and scrap dealers. The gatherer will first return the start-up loan that he received in the morning.

On top of the daily start-up loans, gatherers do not own their carts—an essential tool. Therefore, each gatherer must rent a cart from a scrap dealer every morning. The charge for renting a cart is 5000 TZS (2 USD). At the end of the day, this amount is also deducted on top of the start-up loan. Economically, it does not make sense for gatherers to own one. A new cart can cost up to 100,000 TZS (43 USD). Even though gatherers like Salum, whom I will return to, can afford to own a cart, he regards the gathering job as temporary, with aspirations to get out of the road and become a scrap dealer; therefore, there is no need to invest in them, at least for now.

Scrap gatherers do not enter formal contracts with dealers, which makes it challenging to formalize their labor. A verbal contract arises in the morning when the gatherers rent carts and take start-up money. They are obliged to return after their sales. The amount and worth of scrap materials collected varies daily, depending on the day and where the gatherer went. Some redundancy can happen. A trusted gatherer might not pay for his cart if he had a bad day. However, start-up money must always be returned in full. To supervise the start-up money's return, one of the scrap dealers or an employee is always present at the workshop in the afternoons when gatherers return.

Gatherers' and scrap dealers' relations operate in complete market logic. The relationship between them arises in the morning when the gatherer leases the pushcart and takes cash to buy scrap. The gatherer can decide to go to any of the dealers in the morning and work for that dealer. Upon return, selling the scrap to that dealer, returning the cart, and the working contract ends there. Even though scrap dealers own the tool of production, the pushcart, the control over gatherers' labor starts when the gatherer takes the cart and ends in the evening upon return.

On the same matter, because of their ownership of crafting tools, master crafters can decide who can work or when to take a break. However, it is vital to note that the master crafter-apprentice relationship is not parasitic. They work shoulder to shoulder to produce depending on one's skills, and the labor is awarded accordingly. The apprentice and master crafter have a rather verbal binding contract. Of course, the apprentice is expected to come to the workshop and, in the process of learning, work. However, they have a degree of freedom regarding when to work. In some instances, some did not show up to work for some days. Nevertheless, this is tolerated to certain degrees.

The master crafters compensate the apprentices for their labor. Labor is rewarded for each apprentice depending on the type and number of stoves to be produced. For example, the cost of producing a big charcoal BBQ stove for a restaurant is 80,000 TZS (35 USD). This cost includes aluminum boards, metal coils, welding, and paint. George sells these stoves for 100,000 TZS (43 USD). The marginal profit is 20,000 TZS (9 USD). The apprentice will receive a cut of up to 5,000 TZS (2 USD). The price changes depending on the number of stoves a customer orders. For orders that exceed five stoves, the selling price is reduced by 5,000 TZS (2 USD). If apprentices work together, meaning producing the same stove in a team, each one who participates shares a cut of 5,000 TZS (2 USD) per stove.

Remuneration for each apprentice's labor is based on their contribution to the production process. Unlike wage labor, where compensation is based on the time that one spends in production, here it is the labor itself that is remunerated. Paying for labor rather than the time spent in production reduces the supervision that the master crafter must give to his apprentice. The more one works, the better the pay. For example, Juma and Dinho do all sorts of things, from collecting on the streets, doing some crafting assignments for the master crafters, and sometimes helping scrap dealers load scraps onto trucks or with dismantling and mining metals. They get paid per task.

Hierarchy and specialization are more pronounced in the workshop when examining relations and assemblages around tools. Simple tools like the hammer and *tindo* (wedger) are usually shared without restriction. Advanced tools are tied up with different relations. George charges others to use his circular saw. The circular saw is expensive to acquire and maintain, leading to specific relations between the owner and others. These relationships reveal that the more advanced the tools, the greater the changes in assemblages of labor relations. It also explains the reasons why wealthy people or big companies hire other people to work for them. If George invests even more in advanced tools, certain class relationships will also emerge.

Figure 4.2 Crafters and Their Apprentices Crafting Charcoal Stoves. Photo provided by author.

COMPETITION, CONFLICTS, AND INGENUITY

Even though the workers at Mahakama Ya Friji share ethnic backgrounds or come from same areas, because of scarce tools, unstable market for their products, and customer competition, conflicts become unavoidable. To reduce conflicts that arise from competition, the master crafters have adopted different types of cookstoves that distinguish them from one another. I will return to this in chapter 6. For example, George has started making gas BBQ stoves, distinct from the charcoal stoves made by other crafters. Still, conflicts and competition among workers are everyday matters that the master crafters must deal with, especially Babu Samwel, because of the respect he enjoys due to his age. Conflicts arise from business competition among the master crafters, work competition among juniors, lack of respect, and scrambles for access to vital tools like the welding machine. Although the workers have good relationships with each other, they do not work together in a romantic, peaceful gathering. Because of these conflicts, they have agreed to a rotating leadership among the master crafters.

When I met George for the first time, although we discovered our own Tabora connections, he could not just accept me into the workshop to work with him. I had to explain my research again to everybody. The leadership had to agree. The leadership is also responsible for resolving internal misunderstandings, managing business affairs, and representing the workshop. One of the reasons for Hassan leaving the workshop for a job was a conflict he had with George. Upon his request to return, Babu Samwel convened a meeting of all the master crafters. They had to discuss the matter, but first, they had to resolve the former conflict.

Hassan and George had attended the same class during primary school in Tabora. George admitted that the old squabble no longer mattered and that he would be happy to have Hassan back. Everybody agreed. However, Babu Samwel wanted Hassan to pay *masumule*. Masumule is a long-practiced custom among the Nyamwezi and is paid when one member of the community shows disrespect to another. Masumule can be a certain number of crops, alcohol, or an animal like a goat (Abrahams 1981).

To settle Masumule, Hassan bought a bottle of *Konyagi*, a favored Tanzanian gin. The bottle was shared by everyone at the workshop to show that the conflict had ended. This customary way of solving conflicts holds the crafters together. Their shared customs and values are embedded in the ways of organizing work and resolving conflicts.

DISCARDS LABOR

Discards gathering is not a favorable job in Dar es Salaam. It involves walking around covered in dust and sweat while pushing a trolley full of scrap. Discards gathering is not just at the bottom of the hierarchy; it is also neglected compared to other jobs like working in a factory, street vending, or office work. Nevertheless, neglect is not a major challenge compared to the loathing and violence of those who look down upon collectors and state apparatuses. When electronic discards are exported to developing countries, it is not only hazardous compounds that are exported but also the violence embedded in its fabrics.

I explained the verb "kuokota" (to gather/collect) in the introduction. The noun "*muokota*" is singular, and "*waokota*" plural comes from the same verb. Using the noun "muokota" or "waokota" alone does not make much sense without a second noun, just like the noun fundi, indicating the kind of materials the collector gathers. For instance, "*waokota makopo*," which means bottle or can collectors/gatherers. Scrap gatherers in Dar es Salaam are called "*waokota skrepa*," which means scrap collectors/gatherers. At the same time, "kuokota makopo" (gathering cans) can be used to refer to someone with a mental illness. In that sense, discards gatherers, in certain instances, are reduced to those who suffer from mental illness or abuse drugs. Mental illness stigma toward discards workers have also been observed in Kampala, Uganda (Doherty 2021).

My walimu endure violence and stigma, let alone exposure to hazardous substances. They are considered dirty, outcasts, impure, and contaminated in the eyes of many (Douglas 2003). Nevertheless, their labor products are essential in both local sociocultural activities and the global capitalist system. Their suffering, exposure, and labor are part and parcel of value production. However, they are left unrecognized or uncared for. While salvage labor produces value needed to sustain society and capitalism, both turn a blind eye to the violence that accompanies value extraction from electronic discards (Tsing 2015b).

Crows

On one of my collecting trips with Rama around Magomeni, one of the old neighborhoods created for African laborers near the end of British colonialism in Tanzania, I experienced such violence firsthand. Magomeni is located 3 km from the city center, separated from it by the Msimbazi Valley through which the Msimbazi River flows. Magomeni was known as Mzimuni. Mzimuni was created by the Germans for rich Africans who were former slave owners and involved in the ivory business (Dar es Salaam City Council n.d.).

The idea was to keep Africans away from European and Indian residences while still getting the workforce needed to the city center. After independence, the government—through the National Housing Corporation (NHC)—aspired to build a modern neighborhood, and the area was replanned and affordable housing constructed (Smiley 2013). The neighborhood has since decayed, its infrastructure has collapsed, and the houses are in bad condition. The NHC went bankrupt, and, during the structural adjustment program, the homes were sold to private owners who failed to manage and maintain the infrastructure, let alone the houses.

However, recently the area has attracted wealthy people to invest due to its proximity to the city center. Modern gated homes and apartment complexes are popping up here and there while the rest of the houses display remnants of the heyday of early postcolonial development. "*Angalieni kunguru wa Zanzibar wanapita*" (attention, the Zanzibar crows are passing), I hear a female voice warning. We walk past the house, and still, I don't comprehend. I ask Rama what it means. "Don't you know the Zanzibar crows?" he asks. Of course, I know the Zanzibar crows. "We are like them," he says curtly. We pad along for some time, then he changes the topic, and we start talking about football.

The presence of "*kunguru wa Zanzibar*" (Zanzibar crows) in Dar es Salaam does not go unnoticed. "The skies of Dar es Salaam are alive with the beating of millions of wings"; this is the beginning of an article in *Quartz* magazine about house crows in Dar es Salaam (Teagle 2018). House crows, known scientifically as *Corvus splendens* and in Tanzania as kunguru wa Zanzibar, were introduced into Zanzibar in 1857 by Captain Ward, a British colonial administrator (Vincent 1937).

The house crows are known for their aggressive scavenging of human organic discards. They are also famous for their intelligence—they can make tools, solve complex problems, recognize human faces, and hold grudges (Van Dooren 2019). In 1917, the colonial government declared house crows a pest. Since then, efforts to eradicate them have failed. Now, house crows dominate the East African coastline. These birds are considered scavengers, like the collectors who live off discards.

The question arises, why do they (scrap collectors) bother to collect in areas where they don't find that much and where they experience stigma? Later, as I spent more time with my walimu, I learned about acts of violence that they regularly experience in their endeavors. These are not limited to specific neighborhoods, though they are experienced differently depending on the neighborhood. As argued earlier, discards workers often occupy the bottom of the social hierarchy; they are people who live off detritus, the discards gatherers of modern times and are often regarded as being unfit in society. This is not only the case in Tanzania. Elsewhere, discards gatherers are regarded as outcasts, *catadores*, or untouchables (Bauman 2013; Doron and Jeffrey 2018; Millar 2018).

Thieves

> We have work challenges, it happens, like the other day, I went to take body [a car body]. It was okay, but he did not have the car card [official government car registration], but he is a real owner, so, we signed an agreement, I was given papers, and I was told in case you get caught anywhere, show this, and even my number is there to call, So, from there . . . It was so late, like 8 p.m., before they call for a car lift to carry it in a Fuso, I don't know . . . Isuzu, because I had two bodies. When we arrived here [at Mahakama Ya Friji], I think someone I couldn't know called police officers, and they came. [Imitating a policeman:] "Oya! How is this? Where is a card?" I asked him, "a card?" I showed them its papers. "This is how cars get stolen. You dismantle it there. Where is its card?" This is a scrap, not a car. It is a scrap, this car was just kept inside, and it doesn't work. It is just a body. No, I don't know what, they persisted, persisting was like, come with us . . . "We are taking it to Oysterbay." I don't know what to do. I told him, now, my brother, if they take it to Oysterbay, it is gone. Now they want money! I look at them! So, they took three hundred thousand! That was a profit, Enhee! But thank God that that one made like six hundred thousand. (Rama, personal communication)

Scenes like these are examples of the daily harassment that collectors face in their labor. Not only Rama, but every single collector has also been accused of stealing at some point. One morning, Dinho (another collector) was at the workshop earlier than everybody else. Typically, Dinho is late because he stays up late in the evenings after work, drinking while waiting to help Dada close her grocery, which is attached to the workshop. Dada, as we all call her, lives nearby and owns a bar with a small pork kitchen. The kitchen serves pork in two dishes; one is deep-fried, and the other is roasted pork mixed with vegetables. The two meals are served either with ugali or plantains. On top, there is a salad of fresh onions and tomatoes, thinly sliced or chopped and marinated with lemon juice and a sprinkle of habanero peppers. The bar is usually full of customers who come for food and drinks in the evenings.

We all have a few drinks almost every evening after work. After a long day of working for collectors or hammering in the workshop, we meet at Mangi's kiosk, adjacent to Dada's grocery. Mangi is a common male name for a person who owns a kiosk and originates from Mount Kilimanjaro's slopes. Not every kiosk owner is from Kilimanjaro. Still, generally, people refer to them as Mangi, the renowned Chaga businesspeople, for their entrepreneurial abilities to sense, select, shape, and synchronize the pursuit of opportunities

(Mashenene 2019). Mangi has placed several benches in front of the kiosk that we sit on, sharing a bottle of Konyagi, chatting, laughing, and joking before starting to find our way home, one after another.

We are all used to Dinho arriving late, but not today. Everybody asks why he has arrived so early. "*Nimelala Oysterbay*" (I spent the night at Oysterbay police station), Dinho mumbles. Everybody laughs before asking him what happened. I ask Salum why they are laughing about such a serious matter, as it is not funny to me. "It is his turn. You know, in our work, we all get arrested now and then. People think we are drug addicts; therefore, we steal for drugs. We even sometimes get beaten or called names, yaa such things are like accidents at work."

A security guard had arrested Dinho in Oysterbay. He was accused of stealing the scraps he had gathered in the area. Of course, they let him go the following day because there was no evidence to keep him. However, he has lost the advance money he got from the scrap dealer. In Tanzania, there is a common saying that getting locked up in a police station is free, but you must pay your way out. He is not the first—almost every scrap gatherer has been arrested at least once, accused of stealing. These acts happen more frequently in wealthy areas compared to low-income areas, as we can see in the first incident. Accusations of stealing are even more scary when considering that extrajudicial killings of thieves still happen in Tanzania (Sherrington 2007). Even though such incidents have not happened to any of my walimu, they do still happen on the streets of Dar es Salaam.

Discards Are for Junkies

On a hot, humid day in May 2019, the *masika* (rainy season) is ending after rains have poured heavily all over the city for almost three months. Salum and I have been on the road for six hours. The sun is burning. We are tired and thirsty. Nevertheless, we are in good spirits. Our trolleys are full of scraps. We decide to take a rest under a tree before proceeding to walk back to the workshop. Three young ladies approach us, asking for a particular office in the neighborhood. They are wearing formal outfits with brown envelopes in their hands. I can tell they are going for a job interview. Luckily, we passed by the office earlier on our way. We provide directions and lie down under a beautiful flamboyant tree (*Delonix regia*). An hour later, after we have woken up ready to start the walk back, the ladies appear. I cracked a joke, asking if they got the job. They laugh and ask me how I could tell. We chat for a bit; I tell them about my research. One of them comments, "*tulifikiri nyie ni mateja*" (we thought you were junkies). Now Salum is smiling while staring at my astonished face.

Mteja[1] is a generic Swahili noun for a frequent customer, or in jargon means a drug addict or a junkie. Mwananyamala, where the workshop is situated, was an epicenter of the opioid crisis from the early 2000s to the mid-2010s. It is speculated that the drugs, mainly cocaine, heroin, and crystal meth, were imported from Pakistan and Afghanistan to Tanzania. Kinondoni was the place where the addiction epidemic could be witnessed. When the opioid crisis hit Dar es Salaam, I was a bachelor's student at the University of Dar es Salaam; there were more discards pickers on the street than you could imagine, many of them drug addicts. They would pick anything they could make money out of, and sometimes stole. The opioid crisis increased the crime rate, and discards collection was taken over by addicts (DeutscheWelle 2017).

The stigma of this time continues to this day. Whoever is gathering discards is perceived as an addict and a potential thief. During gathering routines, countless times I heard people calling us wateja. I ponder my walimu's experiences with societal stigma and how it is experienced in different parts of the city. On the Msasani Peninsula, as I explained above, my walimu experienced arrest and detention by police and security officers. These experiences are different in Magomeni and Kinondoni, where verbal and sometimes physical violence is directed toward discards gatherers. Regardless, the reasoning for these kinds of violence is always the same: gatherers are drug addicts, so they either have stolen or are going to steal something.

The societal narrative is straightforward here: crows are tricksters, and they pick and snap things from people, which are the main characteristics of Zanzibar crows. The same applies to drug addicts. They will do anything to feed their addiction. Hence, they steal anything that they think will give them money. And, of course, the drug epidemic produced a considerable number of addict gatherers. It seems that people do not consider the role of discards gatherers filling the gap in the city's infrastructure and addressing the inability of authorities to collect discards, and that their work turns discards into value, producing livelihoods.

However, discards workers themselves see the value of their own labor (as do their families, friends, and associates). Indeed, discards workers see their jobs much like engineers think of engineering or doctors think of therapy. They have a drive and ambition to reach something, like Salum's ambitions to become a scrap dealer, see chapter 5. Regardless of all these conditions and societal perceptions, my walimu expressed this push to prosperity through their labor.

Despite all that discards gatherers offer to the city and the economy, they are still marginalized, not only by society but also by the state. Discards gatherers continue to suffer from harassment by the police and other state apparatuses. At the same time, society continues to stigmatize them. The identification cards introduced by President Magufuli (see chapter 2) would have elevated their status as an important working group in the city, but, as I described earlier, this project was itself scrapped.

DISCARDS CITIZENSHIP

Thieves, crows, and junkies are some of the names that my walimu have to put up with because of their work. The stigma is not only a Tanzanian thing; it is a global phenomenon that has been described by scholars as a form of discards citizenship: designating some citizens as without value. In India, it surrounds lower castes like the Dalit (Dirks 1992; Doron and Jeffrey 2018; Gidwani and Reddy 2011; Harriss-White 2020; Roy 2008), and in China, it applies to the migrant workers who come to big cities from rural areas (Gransow and Daming 2010). It is the same in Vietnam with discards pickers who migrate from rural areas to Hanoi (Nguyen 2018). In Japan, it is applied to the Burakumin (Howell 2016; Neary 1986; Samel 2009) and in Brazil, to the *catedores* (Millar 2018). Most societies have groups of people who make their livelihood from discards. The treatment that these groups receive is not different from that received by my walimu. They face the same stigma of being dirty and impure. They are discriminated against and feared for contaminating others, and governments do not include them in policies and planning. Probably the only difference between my walimu and untouchables in other places is that they do not belong to an ethnic, caste, or racial group in Tanzania.

While society judges discard gatherers because of their labor, Millar finds it hard to categorize their work as informal, done out of desperate poverty. At the discards sites, workers "recreate order by identifying, gathering, sorting and bundling," skills that are the same as in "formal" employment, maybe even more so than office work (Millar 2018). Discards workers on these sites, just like at Mahakama Ya Friji, enjoy the fulfillment of work freedom in relation to wage labor. Many workers, upon finding wage labor, opt to return to the discards site just like Hassan. Gathering discards offers more than an income; freedom of their labor and a sense of community, among other things, reattracted workers back to the workshops. However, outside, they are equated with the violence and drug dealing that happens in society. The condition has been observed in Brazil's *favelas* (Millar 2018) as I experienced in Dar es Salaam.

A similar situation describes Hassan's decision to return to the workshop after two years of working for a wage. I cannot argue that he returned for the joy of working at the workshop. But I would argue that, in part, relations between workers that are based on brotherhood and care for one another were motivating factors, as was the fact that the workshop offers a certain degree of social security. Tanzania's social security arrangement, especially for the self-employed and those employed in small businesses, is not favorable for informal workers. None of the workers at Mahakama Ya Friji are members of any social security fund, and there are no arrangements for their retirement. If things go wrong, however, one can always turn to one's brethren. Some workshop members would travel elsewhere for months and then return to work. They would be accepted immediately, as everyone knows them, and the needed trust has already been established. The brotherhood does not cease once one leaves. I experienced a similar reception when I returned after two years. As I arrived, I was jokingly asked if I would go gathering and whether they should prepare a working station for me to make cookstoves.

Hierarchy, specialization, ethnicity, histories, and networks at play within and beyond the workshop are integral to the interdependent labor process and value circles. Moreover, the workshop is a place for developing and refining lifelong skills. Even though there are rigid divisions and specializations of labor, these divisions are also fluid and plastic. Labor at the workshop seems to be precarious at first glance; however, over time I came to realize that the workshop provides a stable livelihood, as well as a sense of freedom and a place of aspiration for a better life among workers.

Unregulated workshops, like Mahakama Ya Friji, are vital to the city, the country, and globally. They transform discards into immense networks of value and economic activities, let alone the livelihoods they afford to thousands of people. With all this value created through creativity and ingenuity, discards workers experience stigma and violence, sometimes induced by the very state that has failed to build infrastructure for collecting, handling, and disposal of discards.

NOTE

1. Mteja is singular; the plural is mateja or wateja.

Chapter 5

Gathering

Locating and Gathering Discards

Lazima ujue kuthaminisha, hujui tafuta kazi nyingine
(If you do not know how to valuate, do not become a collector)

—Rama, October 2021, Mahakama Ya Friji

The truck arrives at Mkali's scrapyard early in the morning. When I arrive, the loading of scrap iron, copper, and other metals is about to be completed. This is the day that Anton, who works with Mkali dealing scrap, agreed to take me to a smelting factory to sell the scraps collected that week. Several gatherers have been running in and out of the warehouse shirtless, sweating from the humid, morning sun. Dark dust, most likely mixed with refrigerants, electrolytes from car batteries, and probably flame retardants from ashes around the yard, is swirling up. Though the loading job is tough, I join Dinho and the others as they finish loading and tightening the ropes of the tonneau cover. Anton pokes my shoulder to let me know that I will not be allowed to enter the factory. However, he proposed that we should talk to the driver to take me as his helper, a mechanic. I am okay with that. It reminds me of being back in my teen years when I enjoyed helping truck drivers who came to my father's tobacco farm to drive the tobacco to the market.

Mkali is one of the scrap dealers with whom I have been hanging around for four months prior to this day. He allowed me to accompany Anton to the smelting factory on the other side of the city. Anton goes to ask the driver if I can be his mechanic so that I can have access to the factory. "You know he is just a researcher who wants to know where we sell our scraps. That's all, nothing more than that," Anton explains to the driver politely about my research. Only the driver and the seller can drive into the factory with the truck.

After Anton assures the driver that there is nothing to worry about and that I am permitted to drive with them, I jump into the truck between the driver and Anton. We start driving out to the Temeke industrial area. The seat gives me a good view of how the driver meanders around potholes, narrowly avoiding hitting crowds of pedestrians. Aromas of *pilau* and chips *mayai* served by street vendors seep through the open windows of the truck cabin. The air is full of brown clouds of wood shavings from the carpentry shops scattered around the area. We have to stop every couple of meters for other cars to pass. There are groups of mechanics in oily overalls repairing cars along the road. I am told that this old, decayed asphalt road was built in the early 1970s, together with a planned government housing scheme in the area. There are few remains of the project. Although the project houses are surrounded by small businesses and workshops, I can still see them on the line that had been planned. The heyday of Ujamaa and aspirations for modernity can still be observed in this area.

After driving for 20 minutes, we are finally out on the main road and driving south of the city where the industrial area is located. After an hour of being stuck in a traffic jam, we arrive at the factory. The factory is enclosed by an approximately 20-foot-high wall with electric wire on top. There is only one entrance gate equipped with security cameras. The driver honks, and a security guard peeps through a small window and asks Anton and me to disembark. Only the driver is allowed to drive in with the truck. I count six security guards inside the premises, two holding rifles. We register our details in the guest book and are allowed to go in. Anton insists that I stay close to him; otherwise, we would get into trouble. We park our truck at the offloading site.

At one end of the offloading site lies a furnace, with huge red flames that can be seen from where we stand. Sweaty men in protective overalls, footwear, and headgear are dragging scraps to the furnace. Off to the right, around 20 men are busy picking up hot steel rods that come out at the end of the furnace. The rods have been formed from different melted iron materials. After dragging them from the furnace, the rods are watered to cool, then placed in a warehouse or sent directly to customers.

Each site has different and unique ways for how factory workers build relationships with scraps, hot iron porridge, hot rods, and rods that are ready to be shipped out. Interactions with scrap materials involve dragging and throwing. With the hot iron porridge, interactions are more distant and careful. Hot steel rods are held with special gloves, and ready-made rods are more valuable, not simply scraps but new things with value.

The administrative office is separated from the plant, located in a well-painted building with tinted glass windows. After finishing offloading and tallying the weight of all the materials we brought, Anton invited me to a room to observe the day's payment. We sit in a waiting lounge facing a cashier's window. I can see the accountant, a middle-aged man of Indian origin, through the spaces between iron rods, sitting behind a strong iron-enforced window. His hands run over a calculator like a piano player's, and, at the same time, the swirl of a money-counting machine makes the atmosphere in the room dense but jubilant. We are not the only ones in the waiting room; there is another scrap dealer in front of us, and soon the one behind us will also join the wait to receive payment. We have brought with us 3,550 kg of iron (*chuma*), 230 kg of galvanized steel (*dongo*), 96 kg of aluminum (*softi*), 144 kg of copper 8 (*kopa*), 209 kg of cast iron (*casti*), 253 car batteries, 10 kg of brass (*brasi*), 6 kg of stainless steel (*steli*), 14 kg of graphite powder (*fufu*), and three car radiators. The total sale is 2,088,100 TZS (905 USD).

The materials we brought to the factory were gathered from the streets. They were useless and unwanted, matter out of place. They are called all sorts of names—discards, rubbish, or dirt—then are collected, gathered, dismantled, and sorted by gatherers. Upon arriving at the factory, they have gained a value that requires two rifles, a few security guards, and a wall as protection. How did these materials transcend the category of discards to gain this new value? How do the labor of gatherers and dealers at Mahakama Ya Friji recreate the value? This chapter explores these questions, showing that categorizations of discards and value are influenced by the social and political settings of the place and time that an object encounters.

From the beginning of my fieldwork, it was clear that local electronic discards practices (reuse, recycling, repurposing, and mining) depend on global connections. Unregulated recycling practices would not materialize without connections to global economies. On the surface, these connections might seem simple and straightforward: someone in a developed country discards their devices, which end up in Tanzania. That is partially true, but when investigated a bit further, there is interchangeability and reciprocity in the relations between the Global North and South. Whether new or used, devices usually come from the West or Asia. At the same time, some of the materials, for example, metals extracted from electronic discards, enter containers and are exported back to Europe and Asia.

The number of formal electronic discards recycling companies in Tanzania is meager because it is expensive to register and operate an electronic discards recycling company in Tanzania. Apart from registering, the annual fee for an electronic discards recycling company to operate in Tanzania is 10 million TZS (4335 USD), regardless the company size. This fee might seem small; however, considering other costs for salaries, equipment, and establishing the plant, the amount is significant for small-scale dealers like Mkali and others at Mahakama Ya Friji.

Most certified electronic discards companies collect mainly, but not only, from the government and large registered private companies. They do not have the capacity to collect from households, which requires a vast workforce and infrastructure. Apart from this, registered companies would not be able to compete with street gatherers on their prices, especially when buying from individuals. Street gatherers obtain electronic discards through junk-buyer networks at low cost and then sell them to larger companies. This is where the workshop's gatherers step in.

Chilambo General Trade Company, one of the five formally certified recyclers in Tanzania, in collaboration with Ok Plast Ltd., another certified recycler, exports motherboards and batteries collected from the government and big companies to Belgium, France, and Asia for further recycling. They mainly dismantle devices, and parts are shipped off to be recycled abroad. At the same time, toxic parts like batteries and metals that are difficult to extract, such as gold on motherboards, are exported to Europe, mainly to France, Belgium, and Turkey, and to China and India for further recycling and extraction. Based on this, I argue that, regarding electronic discards, there is no recycling plant in Tanzania. Chilambo General Trade Company and others are merely collecting and dismantling places. And this business model will not be able to compete with places like Mahakama Ya Friji.

Mahakama Ya Friji works as a gathering site/container for global electronic discards. The ability of gatherers to reach the vast parts of the city takes over from incapacitated discards infrastructure, utilizing flexible labor arrangements and using simple tools that need less maintenance. Most importantly, the workshop gathers synergies between the formal and informal in response to national politics and reflects the global value chain.

LOCAL GATHERINGS TO GLOBAL MARKETS

Scrap prices, both on the streets and at the factories, respond to local and global markets and politics. Any shocks in the market will affect the price, trickling down to dealers and eventually to street gatherers. During my first fieldwork visit in 2018, prices were at a record low. There were three reasons: First, the global scrap metal market had plunged since 2015, and second, crucially, the new government of President Magufuli had made metal exports difficult because of the *makinikia* (mineral concentrates ores) saga with mining giants. Third, the same regime's policies crashed the construction boom that had been on the rise over the previous 10 years. This meant that demand for metal rods for construction took a steep dive.

Magufuli's regime and nationalistic ideology brought quarrels with investors, companies, and Western countries. He insisted that natural resources and arable land could sustain the country and that the problem was *ubeberu* (imperialism). His politics were anti-Western; to him, the West, on many occasions, meant exploitation. He forced diplomats who criticized him out of the country, leading to the deterioration of foreign relations, and crushed opposition political parties and anyone who criticized him via a surge of charges linked to economic sabotage, disappearing individuals, and going after foreign investors as exploiters of the country's resources (Harris 2021).

One of the most controversial issues was the export of concentrated ores from gold mines. The makinikia saga began in November 2016, when President Magufuli announced more controls over the mining sector to ensure that Tanzania benefits from its natural resources. Two Presidential committees and one from parliament were appointed to review the sector. When their reports came out in March 2017, it was followed by an immediate ban on the export of concentrated ores. Two hundred and 77 containers of concentrate ores from the two largest gold mines owned by Acacia Mining Plc were seized.

The two Presidential reports, popularly known as the Mruma and Osoro committee reports, proved the presence of gold, silver, copper, and sulfur compounds in the concentrated ore containers. The containers held in Dar es Salaam's port were reported to contain raw minerals worth a value of 1.4 trillion TZS that had not been declared. Furthermore, the team reported that the containers held 1.4 kg of gold per ton of ore, exceeding what had been reported by the Tanzania Minerals Audit Agency (TMAA) (Nellist 2017). The Osoro report estimated that between 44,000 and 61,000 containers of gold and copper concentrates had been exported between 1998 and 2017 by Acacia Mining Plc and the former Barrick Gold Corporation. The revealed loss of unpaid taxes was approximately 68.59 trillion and 108.5 trillion TZS (Nellist 2017).

The political stakes were too high upon the submission of the report. The president vividly showed his rage toward the TMAA board and held the minister of energy and minerals, Prof. Sospeter Muhongo, and his deputy accountable. Furthermore, he suspended the TMAA board and the principal secretary for the Ministry of Energy and Minerals for lying to parliament about the concentrates. The report came with repercussions, including the president denying any discussions with the companies unless they paid the arrears established in the report and sought repentance for such acts that had occurred since 1998 (Nellist 2017).

Because of the saga, the minister of the constitution, Prof. Kabudi, presented a bill to the parliament to amend the Mining Act 2010. This resulted in amendments to the amendments of six laws, including those covering mining, petrol, and income tax. Following the passage of the acts, new resolutions were put in place for all matters related to minerals in the country, including disputes, processing of minerals, and smelting of concentrates. The Mining Act required the government to own 50% of shares in mineral firms

Table 5.1 Scrap Selling Prices June 2019 and November 2021 (TZS/kg)

Metals	*June 2019*	*November 2021*
Copper	6,000	13,000
Iron	400	800
Aluminum	2,000	3,000
Galvanized steel	3,000	6,000
Cast iron	900	2,000
Batteries	1,300	1,600
Brass	5,400	6,000
Stainless steel	300	800

Source: Scrap dealers bookkeeping.

and that all earnings be deposited in Tanzania. The TMAA was then replaced by a Mineral Commission appointed by the president. The commission was formed to handle all mineral negotiations and disputes (Nellist 2017). The makinikia saga affected all mineral exports, including those from recycled scrap discards. The price dropped even lower.

When I returned in 2021, one year after Magufuli's death, the country had taken a U-turn in relation to investment. In a matter of a few months after opening the country following Magufuli's demise, the prices rose to almost double what they had been in 2019 (see table 5.1). Of course, prices in the global market were also rising again. The change in the country's politics was a major contributing factor. The workshop and gatherers responded to these changes by adjusting their gathering practices. Gatherers must remain constantly attuned to the ways prices fluctuate so they can adjust strategies. These adjustments are also a form of skill that helps produce value for scrap from below.

SPOORING DISCARDS: THE CITY IN MOTION

After spending two months at Mahakama Ya Friji since my first introduction, I had almost become a member of the workshop. I was invited to evening drinks, and workers had started opening up about their lives, political beliefs, and the football teams they supported. At this stage, I decided to go with gatherers to gather scraps. I was interested in learning: Where do they go gathering? How do they do it? And how is the value determined? Salum and I had become close enough that he was willing to take me on my first trip. Many trips followed after this one, including with other gatherers like Rama, White, and Dinho. Walking with gatherers during our trips gave me a new perspective on the city I had lived in and occasionally claimed as my home.

It is morning in Dar es Salaam, the red sun rays emerge from the Indian Ocean, and the heat rises. I am hurrying to the Mahakama Ya Friji workshop to meet Salum, nicknamed "the Roadmaster" by other scrap gatherers. He is sitting on an old car seat, his back resting against a neem tree; a coffee kettle sits in front of him. He has already drunk three cups. "Sam *Umechelewa!!*" (Sam, you are late!!), he shouts at me. "*Foleni*" (traffic), I reply, giving the usual excuse for being late in Dar es Salaam. We ask one of the *Mama Ntilie*[1] with a food *banda*[2] next to the workshop to hurry breakfast. The neem tree's shadow cools the heat a little bit, allowing us to enjoy a cup of tea, a bowl of beans, and chapati.

I met Salum for the first time in November 2018, when I was introduced to the workshop. In the mornings before work starts and during lunch breaks, there are vibrant debates and discussions about, among other topics, politics, and football. Sometimes these are heated debates, and other times the place is full of joking and laughter. Salum usually sits in the same spot he is in today, smiling at the corners of his lips without participating much in the discussions. It took me several weeks before I started conversing with him. He usually arrives earlier than everyone else, getting to the workshop around 7 a.m. First, he orders his breakfast from Tatu (one of the Mama Ntilie). His order is well-known—a bowl of beans, two chapatis, and a cup of well-sweetened chai. After finishing, he joins the group of others who arrive later. They chat and exchange stories, mostly about their work challenges, successes, and fun things encountered during their gathering trips around the city.

Salum has no problems with anyone. He never argues or fights and is always ready to lend a hand when needed. Several times after our trips, he insisted on sharing the earnings with me, which I found rather uncomfortable. In his early 20s, it is hard to believe that he doesn't spend his money at local nightclubs or bars, unlike some of his peers at the workshop. Now and then, he joins others for their daily evening drink, although he does not drink himself. His lifestyle is motivated by his aspirations to open a scrap-dealing business. "I am going to open my warehouse one day. Maybe 'one day' when you come back next time, you will find me sitting at my place, you will go on the road for me," he told me one day as we took a break under a tree's shade. Other scrap dealers at the workshop all started at the same place where Salum is now. One day means aspiration on Salum's side to move up the hierarchy of the workshop as well as life. Most of the workers at the workshop have aspirations of what life could be; read more in the conclusion. Stories from dealers about how they got capital to start their scrap businesses inspire Salum's hopes that one day he will also own a scrap-dealing warehouse. He knows that prudence and persistence will take him off the "road" (gathering) to a seat in front of his warehouse.

Salum moved to Dar es Salaam in 2017 to join his mother and younger siblings, who had migrated a few years earlier. One of the reasons Salum left his hometown, Kondoa, was to join his family and help his mother pay the school expenses for his siblings. Of course, like many others who have migrated from rural areas, he was also escaping the failure of the agricultural sector and searching for a better life in Dar es Salaam. When Salum arrived in Dar es Salaam, like many, the situation did not turn out as he had hoped. There were no jobs to be found. Hence, he found himself as a scrap gatherer at Mahakama Ya Friji.

Joining the workshop has some complications. Not because it discards work, which would mean that anybody can join. The labor itself is organized around trust, so joining requires a minimal level of trust from the members. Members of the workshop know each other very well. As I have mentioned in the previous chapters, some come from the same region or town, live on the same street, or have crossed paths before. Babu Ali, the scrap dealer, is Salum's maternal uncle. That is how he got in.

At the end of our breakfast, Salum stared at me for a couple of seconds before asking, "Are you sure you want to come with me? It is hard work, and it is rough out there, you know—anything can happen." "Look at me," I reply, showing him my working shorts, a pair of sandals, a t-shirt, a cap, and a bottle of water. He laughs hysterically, pointing toward the trolleys parked in front of Mkali's warehouse and asking me to pick any trolley I like. Later, after experiencing the stigma of discards gathering labor, I understood what Salum meant by saying anything can happen.

When gatherers leave the workshop in the morning, they have to decide which trolley to take, typically depending on where they are going that day. If the day trip focuses on the Msasani Peninsula, they take the lightweight trolleys made with regular bicycle tires because the area has more paved roads than other parts of the city. The area is occupied by Tanzania's wealthiest community, along with politicians, diplomats, high-end hotels, bars, and nightclubs. These trolleys are lighter, and bicycle tires allow for a fast-walking pace. If the collector decides to collect from uswahilini (low-income neighborhoods), which have mostly unpaved roads, the collector will take a trolley with motorbike tires. These are heavier, but because of the width of the tires, they drive smoothly over sand and mud. Salum had decided to spend a few hours in Uswahilini, near Bagamoyo Road, so that we could cross to the peninsula later. For that matter, we picked the lightweight trolley with bicycle wheels.

We start walking out of the workshop. Salum decides to start in the Mwananyamala area, a mix of lower- and middle-income households, instead of his preferred affluent neighborhoods along the Msasani Peninsula. This decision is reached after a thorough evaluation of the day—where other scrap gatherers plan to go, the route used in past days, and potential places that have been observed during previous trips.

"*Huwa sipati sana upande huu. Hata kama nikipata wanaganga sana bei. Lakini tuanzie huku tutaenda Msasani mchana*" (I do not get a lot in this area. If I get anything usually, I have to negotiate a lot for the price). Evaluating the day here means navigating networks and assessing how best to exploit them. This, in turn, means the evaluation is factored into the final service. "We have to start here, and then we will go to Msasani Peninsula in the afternoon," he explains. We walk for a kilometer along the narrow streets of Mwanayamala, turning left and right every couple of meters, meandering around children playing football. "Puff," a football hits our trolley, followed by a scream—"*quiii aahhh mamaaa*" (crying sound)—from one of the children. We pass a couple of houses down the street before taking a turn into a small empty space. "Shak shak," a sound of clothes being hand-washed in a plastic basin. "Quaaa quaaa," another sound most likely coming from cleaning pots with steel wool. We keep walking. At a crossing, I try to jump a puddle but fall short; "chatwaa," my right foot just hits the edge of the puddle, and the water splashes up to my knee. While we laugh about my failure, we hear a voice from a radio: "Radio One, Radio One, stereo." A middle-aged man wearing a Muslim hat, locally known as *baraghashia,* is listening to a local radio channel. He gives us a suspicious glance.

We keep walking and decide to switch on a speaker attached to the trolley with a prerecorded message: "*tunananua skrepa na machine mbali mbali zilizokufa, friji, computer, simu nakadharika*" (we are buying scraps, broken machines such as fridges, computers, telephones, and the like). Discards gatherers like Salum take to the road every morning, roaming the streets with loudspeakers announcing that they will buy broken devices.

The loudspeaker plays a critical role in a gatherer's labor. The loudspeaker alerts people of a collector passing by. However, there are politics to the speaker's use. Gatherers do not switch it on as frequently or loudly in affluent neighborhoods compared to other areas, especially on the Msasani Peninsula. At the peninsula, I suppose that the upper class prefers tranquility, and they can afford to pay companies to collect their discards. Gatherers are meticulous about their use of the loudspeaker. Several times, I heard that someone got in trouble on the peninsula because they were making noise. In uswahilini, however, a collector will get nothing without having the speaker on.

As the gatherer passes by, people bring out their defunct devices. The device *itathamanishwa* (value estimation); based on how much valuable material (metals) is in the particular device, the collector decides on an amount of money to offer, and a negotiation will follow. Usually, the collector offers half of the amount he is willing to pay. The price then depends on the seller's ability to negotiate for higher pay.

Meandering through narrow streets, we often had to jump over rainwater puddles. Back in the days before Mwananyamala was integrated with the city, the valley was full of rice paddies. As gentrification took over the city center, people living on the margins were pushed to the valley. Unplanned housing emerged. The area has poor infrastructure and lacks rainwater trenches or sewage systems to direct rain and sewage water away from people's residences.

To gather discards in the cityscapes is embedded within the broader "arts of noticing" (Tsing 2015). After we hadbeen walking for an hour and have found nothing yet. Salum decides that we should go to the demolition site he saw the previous day. Though the site is only a couple of streets away, several turns result in me losing my bearings. After 20 minutes of walking through congested streets, we finally reached the demolition site. Two old houses are being knocked down. The houses have been bought by a wealthy person who wants to build a villa. As it has become more costly to buy land in the city center, rich people have also started moving to the valley, pushing poor households even further from the city.

The tin roofs are brown, and one can see small beams of light shining through many nail holes, indicating that the roof has been repaired multiple times. The foundation has sunk into the ground because the area is swampy, indicating that the house has survived several floods. I guess that houses in this area were built two or three decades ago. We briefly chat with the guys at the site. They allow us to go in and see if we can find anything. We spot several water pipes, but they are too long to fit in our trolley. We cannot take them with us today. We asked when they will take down the roof so that we can collect the old iron sheets too. Other scrap materials, not limited to electronic discards, are also part of the materials collected. The iron sheets will earn a significant amount of money at the scrap market. We exchange phone numbers, and they promise to call us when the roof is dismantled.

It is just past midday; my stomach starts demanding food, and I am sweating and dehydrated. The 1.5-liter water bottle we bought in the morning is almost empty, and my legs have started to shake. Because gathering involves walking during the day, the sun and heat of the city can dehydrate the body. Every gatherer has to have a bottle of water, sometimes two. Of course, the water bottle is not the most important tool for this problem; it is the unique nylon *rafiki wa mazingira* (friend of the environment) bag that every gatherer has. The bag is usually found hanging on the side of the trolley. Something special about the bag is that it keeps the water bottle steady when walking, accessible, and cool. This bag has no other function than holding the water bottle. The bags are made from recycled plastic. The plastic threads are woven close together, making the bags durable. However, the most important is that when holding a cold water bottle, the moisture on the bottle keeps the bag wet and cool for a long time, regardless of the Dar es Salaam heat; this is why they are preferred.

Salum can see my condition—he suggests we cross Bagamoyo Road, which separates Mwananyamala from the opulent Regent Estate neighborhood. Before we cross, an old lady comes out from her house to say she has a broken microwave. She heard the announcement from our speaker. We go into her backyard to look at it. Salum inspects the microwave keenly. I follow his eyes, searching for any sign of valuable metals, especially copper. He decides to give the lady one thousand shillings (about half a dollar). She gets angry and asks us to leave. As we leave, I ask Salum why he offered that price. He replies, "*Nilikwambia watu huku wanafikiria wanaweza pata pesa kwenye kila kitu, microwaves hazina kitu*" (You know people here think they can make money out of this, there is not that much in microwaves).

Months after this first trip with Salum, when I became more familiar with different areas, I came to realize that although I had lived in this city, I had never really mastered its nitty-gritty. I knew several neighborhoods and famous streets, but I had hardly walked around these or other areas. My knowledge of the city had been limited to bus routes because I traveled by daladala (public buses) most of the time.

Walking during the gathering gave me a different way of knowing the city (Ntapanta 2023). My understanding of the city changed considerably. As Tim Ingold notes, walking is "itself a process of thinking and knowing" (Ingold 2010). Each step I took, each tree shadow, and the street corner I passed not only made me feel the city but also enabled me to become knowledgeable about it (De Certeau and Mayol 1998). I had never before understood the interlocking entanglement of the city, its people, infrastructures, and everything else that makes Dar es Salaam live up to the name "Bongoland" (the land of brains) (Calas 2010). It is said that to live and earn a livelihood in Dar es Salaam, one must use their brain more than their physical capabilities. Even though gathering scraps is very physical, gatherers need to have sharp knowledge as well as physical capabilities.

CROSSING CLASS LINES

We cross over Bagamoyo Road to the Regent Estate, and the scenery changes instantly as we walk toward the peninsula. Bagamoyo Road separates the affluent neighborhoods of the northeastern coast of Dar es Salaam—including the Msasani Peninsula, Masaki, Mikocheni, Mbezi Beach, and others—from densely populated areas like Kinondoni, Mwananyamala, Sinza, and Mwenge. Tall buildings line both sides of Bagamoyo Road. Salum opts to start on Chato Street. At the beginning of Chato Street stands an abandoned villa that used to be the AAR Hospital, separated by a wall from the Sunrise

Primary School, an English-medium private school. Yellow school buses with English alphabets painted on them are parked outside the gate.

Further down the street, there is an old colonial villa known as Mfuruki House, followed by house after house of well-planned order. The majority of the villas have one or two stories, and each house has a wall around it, some with electric fences on top, and a gate facing the street. The roads are paved with water trenches, and trees are planted along the pavements. After the Sunrise School, we turn right onto Urnusio Street, passing the Tanganyika Law Society (TLS) building. TLS is very respectable and is the oldest lawyers' association in Tanzania. After TLS comes the Crown Health Centre, a private hospital; on the other side of the road lies Addis in Dar, an Ethiopian restaurant. Across the Migombani Street junction, 400 m down the street, stands the magnificent Shoppers Plaza where the Fairy Delight bakery is located. In the bakery, one can get a Danish pastry as delicious as those in the Lagkagehuset Bakery in Copenhagen.

I feel a little better walking in the shade, and it seems easier to walk here compared to the previous neighborhood where we had to meander around puddles, stones, and kids playing in the street. I hear a few birds perched on the trees singing beautifully, and very few people can be seen walking. I guess that those we see are people working in the area. In contrast to the previous neighborhood, I do not see many street hawkers pushing trolleys full of merchandise here. In this neighborhood, newspaper stands, fruit vendors, and kiosks are strategically placed at junctions. I feel somehow intimidated by the voice of our speaker. This is because of the difference in environment and the class-induced calm of the area. Most people pass by in cars, beep in front of a gate, and drive in with the gate closing immediately behind them.

However, after two hours in the Regent Estate, we have two washing machines, a fridge, and a treadmill. We buy the washing machines from security guards in two different compounds. One explains that his boss asked him to find a way of disposing of the device. He knew that scrap gatherers would pass by, and he could make some money, so he had been waiting. After hearing our loudspeaker, though playing at a very low volume, he rushed out to meet us. With the second washing machine, the security guard was given money to pay a formal recycling company to collect it. Before he made the call, he heard us passing—which means he earned double. The fridge and treadmill are given to us for free. The owner is driving out from his house and sees us pushing our trolley. "*Nyie, kuja hapa. inaokota vitu mbovu*?" (Hello, come here, are you gathering for broken things?), he asks in Swahili with a heavy Indian accent. He takes us to the garage and shows us the items. "*Chukua hizi inazuia njia*" (take them they are blocking the garage way), he insisted. We call it a day and walk back to the workshop.

Discards are lively in the cityscapes. Extensive infrastructures are set around urban centers to get discards out of sight. Moreover, even when they arrive at dumping sites or incinerators, they continue shaping landscapes, cityscapes, and airscapes. They shape landscapes, disturb the city's ordering, and occupy airscapes. They are mobile, lively, and cityscape doers with and within cities. They are blocking the way, revealing their agency to disturb and shape human organizations of what spaces should be. Therefore, the gathering is a way of describing particular discards' doings and their socialities (du Plessis 2018, 125–126).

CLASS AND DISCARDS

Price negotiations depend on the value of objects and on the seller's knowledge of that value. Price is determined by the knowledge and economic muscle of the seller. Devices that I collected with my walimu during my fieldwork varied from one area to another and across individual sellers. For example, at the Msasani Peninsula, Regent Estate, Mbezi Beach, and Mikocheni areas, the collected devices were in good condition, and we got them for cheap or sometimes for free. The polarization between the rich and poor, similar to the division between the north and south, can be understood through the materials collected. This also tells a lot about repair, who repairs, and ideas about repair (revisit chapter 3).

In low-income neighborhoods, like some in the southwestern part of the city, however, the devices tended to be older and damaged, with visible marks from multiple repairs. Sometimes, parts had already been plucked by repairers, and price negotiations tended to be longer. Valorizing fluctuates across the class. The upper class doesn't ask for much money for sometimes better products. Factoring in class and urban stratification is also part of the gathering job.

Salum prefers to go to the peninsula, as described above. Rama, another collector, takes the opposite direction. He doesn't like to go to the peninsula. First, he is afraid of the many police and security guards who constantly harass gatherers. Second, he does not have a good network on the peninsula compared to Salum. Rama prefers to go to mixed-class and emerging neighborhoods like Sinza and Mwenge and low-income areas like Tandale and Kijitonyama. During our trips, he was more comfortable in those neighborhoods. He would let his recording run throughout the day.

On my trips with Salum to the peninsula, we had to switch our speakers on in certain areas while we could not do so in others, especially in opulent streets. In these areas, people valued their tranquility, hence, Salum depended more heavily on the network he had created with security guards, gardeners, and housekeepers than on the loudspeaker. These people would call him when they had things from their bosses. When I visited the workshop in September 2021, Salum was no longer going for long walks. He would wait for a call, then rent the cart to go collect the items.

Sales and negotiations are influenced by many factors, including the device itself, the neighborhood, pre-existing relations, language (especially the use of specific jargon), who is in the household at the time of the sale, and the selling family's income. In a middle-class neighborhood, negotiations over price are usually brief, and people sometimes prefer to give things away for free as otherwise they would have to pay a company to come collect them. In many cases, it is the opposite in low-income areas. Knowledge of which devices, types, or brands are profitable and how to handle price negotiations is crucial in gathering electronic discards (Ntapanta 2023).

KUTHAMINISHA: VALORIZING ELECTRONIC DISCARDS

On the "road," gathering is not only about knowing where to go or selecting the neighborhood. It also involves *kujua vitu* (knowing the objects), *kuthamanisha*[3] (knowing the value), and *kuganga*[4] (negotiating) a good deal with sellers. When a device is presented to a collector, very quickly, he has to know what kind and weight of metals are in that device. He has to negotiate a buying price with the seller while at the same time keeping in mind the day's price at the scrap dealer. All of these meticulous calculations happen in a matter of a few minutes. This knowledge and these skills have to be transferred through apprenticeship before one can master gathering enough to go solo. In most cases, the value is determined by the type, brand, size, amount, and types of metals embedded in a device. Big electronics like refrigerators and air conditioners have a higher value than small electronics. For my walimu, a good day means gathering, for example, a wall-mounted air conditioner.

Most wall-mounted air conditioners are old and were built for Europe or North America. Wall-mounted air conditioners are no longer as common. Because of technological changes, most air conditioners in use today are split-system air conditioners. Wall- or window-mounted air conditioners consist of one unit attached to the wall (or window), with a fan pumping cold air inside and the heat pumped out from the back. Conversely, split-system air conditioners have two or more units, with the condenser outside the building and connected by a tube to the evaporator inside. The old air conditioner models contain a considerable amount of metal, specifically copper, while newer, advanced air conditioners have less.

One day, I brought an air conditioner that had been discarded in my residential complex to the workshop. The air conditioner was from a Chinese brand, Toyo, which is also famous for its cheap motorcycles. Upon seeing the brand, the scrap gatherers bluntly told me it was not worth much. I wondered why: is it not the same as a Hugo or Samsung? They explained that this brand and several other cheap brands from China don't use much copper in the wiring. And because sometimes wires are copper-plated, that's why they don't last. As one can see, this was not just the size of the item but reflected a more complex valuation system that requires in-depth knowledge about where, when, and for what market a device is produced. This realm of knowledge includes knowing about corporate or multinational industrial design and product manufacturing choices made in mass production.

White, another collector, had recently shifted from crafting charcoal stoves to gathering because of the rise in scrap metal prices. He had a bad month when I visited the workshop for the second time in September 2021. He was approached by a man selling his old Toyota Corolla that had deteriorated beyond repair. The man wanted 800,000 TZS (347 USD) for the car. After quick negotiations, they settled on a price of 600,000 TZS (260 USD). However, White had never bought a scrap car before, and he could not know the car's weight because he could not weigh it.

A simple technique that gatherers use, once they know or guess the types of metal embedded in an object, is to lift it up. The purpose of lifting is to estimate the object's weight, which provides insight into how heavy the metals are. This technique becomes useless with big scrap items like the car that White was offered. Therefore, the collector must gamble on the lowest price possible when there is no precise value estimation.

I discussed the deal with Rama, and he bet that White would end up with a loss. He told me that right then before the deal was concluded. A Toyota Corolla should weigh less than half a ton of iron, and with the engine, the tires, and the copper from the wiring, Rama predicted that White might get between 400,000 and 450,000 TZS (217 USD).

However, White got 500,000 TZS (217 USD) with a loss of 100,000 TZS (43 USD). At the time of the sale, the price for iron was 800 TZS per kg. Five hundred kilograms would be worth 400,000 TZS (173 USD) on the iron alone, plus other materials and the car engine. But White had miscalculated, and the engine was not sold. Until I left the field, White was still waiting for a repairer who was supposed to buy the engine.

Then Rama told me the story of when he was approached by a car mechanic selling a defunct Ford Ranger truck. The mechanic wanted 400,000 TZS (173 USD) for it. Rama proposed 200,000 TZS (87 USD) for the car as his initial price, telling the man that cars would not give him anything for their price. One of his explanations was that the truck didn't have its gearbox, making it less value to scrap dealers. However, he had spotted that it did have its complete electric wiring, which he estimated weighed around 10 kg of copper.

The car's chassis was still intact, and the springs were there. Sharply, he estimated that the car would weigh around 800 kg, making its body worth around 640,000 TZS (277 USD). With other metals, he reached a total estimate of around 800,000 TZS (347 USD). He then had to calculate costs for the labor of dismantling the car, transport from the garage to the workshop, and renting a circular saw for cutting; these costs were nearly 100,000 TZS (43 USD). All of these calculations were made on the spot while negotiating the price. In the end, they settled for 350,000 TZS (152 USD), which made Rama a profit of 350,000 TZS (152 USD), equal to a month's salary for a new primary school teacher.

White had not yet developed the skills and knowledge that Rama and others hold about which devices, types, or brands are profitable and how to handle price negotiations. White had not been gathering long enough to navigate the car negotiation; these skills require a long apprenticeship. To know the gathering business, one has to master complex knowledge, and this

Figure 5.1 Collecting in Mwananyamala with Rama. Photo provided by the author.

knowledge is learned through trial and error (Lave 1977). Rama has mastered the gathering craft after working for more than five years.

When a new member joins (usually a man, as there have been no women working as scrap gatherers at the workshop), he has to go gathering with an experienced member. The rookie will observe every step, from where to go to which devices are most profitable, and will learn to keep up to date with price fluctuations from dealers that guide price negotiations with sellers. All this knowledge has to be tailored to the apprentice, hoping that he will master it after a couple of trips. When dealers are satisfied with the performance of an apprentice, then they can borrow the start-up amount and go out on their own. White will consider this the next time he buys cars.

DEALING DISCARDS

On my first day gathering, before leaving the workshop, Salum walks toward Mkali's warehouse. Mkali is sitting by the doorstep on an old safe. He hands us 50,000 TZS (22 USD) while joking that Salum should look after me so that I don't abscond with the money. Mkali, like Babu Ali and other dealers, is an intermediary between discards pickers and big scrap companies. They buy scrap and metal from street gatherers. Discards gatherers gather, dismantle, and sort the metals to sell to the dealers. The amount Salum borrows from Mkali will be deducted in the evening after the sale. Whatever remains after paying back the advance and the 5,000 TZS (2 USD) for renting the trolley is Salum's income for the day.

Several tools can be found at dealers' warehouses. The essential tool for any scrap dealer is the scale for measuring weight. The whole scrap business is about the scale and the record books. Everything bought and sold from gatherers is measured and valued by weight. The value of scraps and metals is measured by weight. During weighing, the weight of the materials is written in a counter book next to their price. A scrap dealer cannot work without a pen and a book. All transactions have to be precisely jotted down to determine the value and keep track of the transactions.

The books are archives of gatherers' labor and dealers' business and are valuable for researchers like me. However, the book and the pen are also significant regarding power and hierarchy at the workshop. The pen and the record book are symbols of that hierarchy and wealth. While crafters and gatherers do not keep records of their expenditures and income, scrap dealers are meticulous and very serious about keeping records. Babu Ali has kept his book for the past five years—one counterbook for every month. The details of these books include the date, every collector who received the start-up money, how much, if they rented the trolleys, what materials were collected, by whom, how much every collector received on that day and finally, all the sales to the factories.

Babu Ali gave me two reasons why he keeps his books that way. First, he told me a Swahili proverb—"*mali bila daftari hupotea bila Habari,*" meaning "without keeping records, wealth disappears into thin air." It is essential to keep track of expenditures and income for a scrap dealer to stay in business. Second, scrap dealers are registered with the tax authority. These books are vital for estimating how much taxes they have to pay. Here is the junction where the formal and informal meet. Informal labor is never pure in the informal sector but is interwoven between formal and informal regimes.

Once a dealer has collected enough—usually between 2 and 10 tons—scraps are sold to local factories for production or are exported abroad. Scrap dealers mentioned a few well-known companies, though it was impossible to access their facilities or talk to them. Scrap dealers identified the companies by the ethnicity of their owners: *Kwamchina* (the Chinese), *kwamlebanoni* (the Lebanese), *kwampalestina* (the Palestinians), and *kwamuhindi* (the Indians). The factory where I accompanied Anton was kwamuhindi. Scrap dealers' choice of where to sell their scrap depends on where the price is higher on a particular day, which also depends on price fluctuations in the market. Once a dealer's warehouse is full, phone calls are made to determine where the price is highest.

Dealers at Mahakama Ya Friji and others nearby usually agree on a uniform buying price for scraps from gatherers. The buying price set by the scrap dealers is written on wooden signboards standing out in front of the warehouse. The price depends on the selling price at the kwamchina, kwamlebanoni, kwampalestina, and kwamuhindi factories that export or produce objects out of scraps. However, on my return in October 2021, a new dealer had opened a warehouse not far from Mahakama Ya Friji. One day, I passed through the new big iron sheet building to discover that his buying prices were higher than the other dealers. I asked Babu Ali why the price there was higher than theirs. "We are also wondering how he can run the business with that price. It doesn't reflect the market price. His profit should be very marginal," he wondered. However, Salum had a different perspective. He thought that the new dealer was trying to put others out of business to build a monopoly. Salum insisted that he would not move to the new dealer, regardless of the higher price: "As far as no one is dominant, it is better for us gatherers."

CONCLUSION

On our last gathering trip, I asked Rama why the price now, in 2021, was 14,000 TZS (6 USD) per kg of copper, astonishingly high compared to 6,000

TZS (3 USD) when I first arrived in 2019. "*Unajua baba alipokuwepo alikua anasema visitoke nje, sasa mama kafungua ndio maana*" (When father [referring to Magufuli] was around, he forbids exports abroad, now mother [referring to President Samia] has opened, that is why the prices have gone up), he replied. But the scarcity of scrap needed for the smelting industry was also a factor in pushing the price up.

The scrap business reflects and absorbs global price fluctuations and national politics. The makinikia saga had a tremendous impact on the business. Scrap dealers create networks that organize the scrap business. On a closing note, the networks that scrap dealers create with gatherers by providing tools and start-up funds aid the transition of scrap materials from houses and the city to the workshop, and then upward to companies that buy scrap in bulk for manufacturing and export. Local politics become essential in shaping the policies that affect the scrap business.

Gathering discards is an art that involves skill and expertise in knowing and extracting value, knowing the history and provenance of discarded goods, and understanding the way the city is arranged by neighborhoods and income, providing a roadmap of where and when to go. In negotiating with device owners, a discards gatherer has to project the value depending on the type and size of the device and what he is going to earn from it. On top of that, he has to lower the seller's expectations to agree with his price. At this point, the discards gatherer needs to know the types of metals he can find and what price he will sell them at. These are detailed forms of knowledge to acquire.

To gather is to observe the city, the workshop, the tools used, the objects (cookstoves) produced, and the scrap business. These assemblages of relations connect to each other and to more extensive networks beyond the workshop, the city, and the country. Market shocks, for example, or changes in national politics and policies are felt all the way down to the workshop level. This chapter has connected these levels, not in hierarchical order, but as webs of relations, labor, and tools that are important to understanding electronic discards recycling in Tanzania.

NOTES

1. Women who sell food.
2. Makeshift buildings, usually made of cardboard or iron sheets, where food is served.
3. This term is widely used when estimating the value of objects, either by comparing or inspecting for faults or perfection that might lower or raise the value.
4. A colloquial word used to mean stubbornness in getting something.

Chapter 6

Crafting Discards

Tools, Knowledges, and Imaginaries

Tools are the juncture of crafters' bodies, skills, and knowledge with material objects. Tools facilitate the worker's art and actions; they translate thoughts into actions. At the workshop, the deployment of tools allows cookstoves to be made, materials to be transformed, and labor to be simplified. In this chapter, I explore the gathering of tools in crafting and the ways they are negotiated and express their agency in craftsmanship. The chapter extends into labor relations and how specialization arises in relation to skills, imaginaries, and materiality.

The master crafters and their apprentices form different strata of labor relations. The apprentices, usually young, come to learn the art of crafting with different aspirations. Juma, who lives with his mother just a few streets away, wants to stay and work at the workshop. The younger Samwel wants to learn the craft and, at some point, move somewhere to start his own place. The apprentices have unwritten contracts with the master crafters; however, while unwritten, these contracts are well respected. This hierarchy is important in the division of labor and production processes. Because there are no fixed prices for the charcoal stoves, prices are always negotiated depending on the needs of the customer. Apprentices have not yet grasped the nitty-gritty of the production costs, which materials are preferred, designs and selling prices, so it is for master crafters to control the production process.

Apprentices are expected to show up at work when there is a job to do. However, when there is no work, they are not required to be there. At the same time, having apprentices allows master crafters to engage in more ventures. George does business with restaurants, making stoves that use gas, which allows him to work flexibly outside the workshop. If the master crafter has received a large order for stoves, he may assign his apprentices to work on different parts of the stoves. However, when there are several small orders, the master crafter might assign each apprentice to work on one order from beginning to end. After instructing an apprentice on what to do, he can go around the city to repair and fix stoves while things run smoothly back at the workshop.

The master crafters all started as apprentices early in their teen years. The crafters need to have skills including measuring, cutting, stitching, and composing materials. These skills do not evolve from mere intuition; crafting is a vocation (Weber 1958). George and the others have worked on making things for years. Skill is a trained practice as opposed to *coup de foudre* or mere inspiration (Sennett 2009). These years of skill-building allow the master crafters to specialize in different types of crafting, which I describe later in this chapter.

Even with these specializations, however, the crafters share a common trait: material consciousness. Material consciousness makes objects interesting, arousing, and allowing the crafter to turn their thoughts into reality (Sennett 2009). Material consciousness manifests as a two-way traffic. On the one hand, a crafter thinks about an object that he wants to craft, then he thinks of suitable materials. Inspection of the materials includes touching, gazing, and pressing of refrigerators and air conditioners is not a search for aesthetics but rather an assessment of the quality of materials. On the other hand, after seeing materials, the crafter begins inventing how and into what these materials can be shaped. This particular creative process happens when collectors return to the workshop in the afternoon. Crafters salvage what the collectors have gathered, either by searching for specific materials they want to use or by looking at the available materials to spark their minds in terms of what they might be turned into.

These skills are developed through long apprenticeships and skills development. However, as I have mentioned previously, certain prior relationships are necessary to be accepted to learn the art of crafting at the workshop. One cannot just pay their way in. There is a thorough social vetting that requires an individual to meet certain criteria. This process is quite the opposite of the easy entry feature of informal labor.

CRAFTING RELATIONS

While waiting for George on my first day at the workshop, as I observed Mjomba Mjomba working on a cookstove, I had to be patient and not interrupt his business. When George had time to talk to me, he invited me to sit with him in front of one of the kiosks. "So what do you actually want to do here?" he asked with curiosity. I explained my research and my interest in following how and what they do with defunct electronics. He asked me where

I was from in Tanzania and which kabila I belonged to. "From Tabora," I answered. I could see in his eyes a confirmation that I didn't understand. "I knew," he replied. He was also from Tabora. My surname is easily recognizable among the Nyamwezi people. He explained to me that most of the crafters are also from Tabora. Now that our conversation had become more straightforward, we spoke at length.

George's father had been a correctional officer in Tabora (as had my step-grandfather and my father). Though George's father was not a Nyamwezi, unlike my step-grandfather, his mother is Nyamwezi, so he became one too. Early anthropologists of Nyamwezi society observed that Nyamwezi kinship relations did not follow particular anthropological labels like patrilineal, matrilineal, or cognatic. The Nyamwezi system could be described as a combination of a range of structural and behavioral features (Abrahams 1981; Bennett 1971; Rockel 2000; Unomah 1977).

Anyone can assimilate and become a Nyamwezi. The famous Nyamwezi military segment "*rugaruga*" was made up of Nguni warriors who had migrated from southern Africa to settle among the Nyamwezi. George's father was originally a Luguru from the Uluguru mountains near Morogoro. However, George could become Nyamwezi because of his mother's lineage, in a way similar to me: I am also Nyamwezi through my grandmother's marriage to a Nyamwezi, my step-grandfather. Although my grandmother originated from another group, the Gogo, I am also considered Nyamwezi and was given my step-grandfather's first and last name. It turned out that George had heard about my grandfather from his father, and I also knew his father's name, though I never met him. Suddenly, in our first chat, we discovered that we were connected back through Tabora.

Nyamwezi are known for splendid etiquette, which reveals itself in their customary greetings. It is considered offensive to start a business discussion without a lengthy greeting that includes discussions of one's health, journey, family, properties, and other things (Abrahams 1981). At the end of my grandfather's life, before he died at 94 years old, I remember visitors coming to visit him at his farm just outside of Tabora town. The visitors did not state the purpose of their visits until food was served. And, if they were important visitors, a chicken was slaughtered. After everybody had eaten and was refreshed, my grandfather would invite them to sit under the big mango tree. There, the business talk would start.

I felt the same thing when George took me out of the workshop to sit on benches in front of the kiosk. He asked me if I would like a drink. Awkwardly, I said I would pay for our drinks, as I had interrupted his work. He refused my offer, which would be *kaponda soni,* a shame for the visitor to pay. We had a beer together and discovered all I described above. Later, George introduced me to the leadership as "*bandugu yetu*" (our relative), and among the rest of the members, I even got called "*wang wise*" (a person from our motherland).

Even though crafters came from the same area, rarely did I hear that their families had any connection. The crafters became friends for years and were regarded as kin to each other, so it would make sense that their wives and children might have strong ties too. But that didn't seem to be the case. Over all the eight months that I spent at the workshop, only George invited me to his house, and only he visited mine. Ethnicity tends to be reconfigured in urban centers where domestic life is separated from the economic sphere, even between members of the same ethnicity working at the same place. On top of that, because intramarriage is not prevalent in Dar es Salaam, most of the workers are married to people from other groups, curtailing the ties between domestic spaces (Cohen 2014).

Ethnic specialization, especially in informal economic activities, is common when people from the same ethnic background migrate to urban centers (Parkin 1969). However, I would not argue that sharing ethnicity determines one's acceptance at the workshop, but rather that relationships, like coming from the same village or town or knowing somebody at the workshop, determine acceptance. Sharing an ethnic background is secondary. Workshop members accepted me in a similar way to how relationships are formed among the workers, and to how the workshops themselves are formed as social collectives. These relationships, which according to Abraham (1981), are embedded in the Nyamwezi, become more of kinship relations than being oriented solely around ethnicity. These kinships-making connect the workshop beyond Dar es Salaam to rural areas like Tabora. The workshop is connected to values, skills, and knowledges from broader ethnic-cultural and technological landscapes.

TOOLS AND CRAFTING

Tools are central for any crafter to engage with the thoughts, skills, and materials needed to produce objects (Ingold 1997). Tools are connectors between the mind, the body, and materials. However, when tools are discussed, they appear as a means to an end, used to produce or shape something, not an end in themselves (Graham and Thrift 2007; Henke 1999; Harper 1987).

The tools at the workshop are of three general types: First, tools that are co-opted to perform different or more roles than those they were intended for, like the rail. The rail was not intended to be used by crafters at the workshop but has been acquired to perform roles beyond enabling locomotives to travel on it. On the rail, electronic discards are dismantled, shaped, and smoothened. And crafting is happening. The noise I heard on my first day at the workshop was the hammering of material on the rail. Owning a rail means that the crafters are able to work at any time and for as long as needed. Borrowing it from others can lead to delays when the owner wants to work with the tool. However, apprentices have the privilege of using their master's rail when they are working for him.

The second set of tools is those made by the workers to execute specific tasks. This category comprises tools made at the workshop, including the bender—an imitation of an industrial tool that has been made by crafters using scrap materials. The last set of tools is the modern, industrial electric tools, like the circular saw, that workers have bought. These tools are acquired, bought, converted, or made depending on the needs of crafters and their affordability.

Essential tools that crafters, especially master crafters, should have at *Mahakama Ya Friji* include the rail, a *tindo*-wedge-shaped iron rod, a car seat, and a hammer. The toolboxes comprise all small tools like scissors, screwdrivers, and others. Owning these vital tools makes a crafter independent. It takes a long time for the crafter to gather the necessary tools. Only the master crafters possess most of the tools. The apprentices start collecting tools as they learn, and once graduating, the have acquired most of the tools necessary for the work. Of course, workers borrow tools from each other all the time. However, once the owner wants it back, it must return it.

Possession of vital tools creates particular power relations among the crafters, as I described in chapter 4. Some of the master crafters, like George and Mwanawane, are investing in expensive electric industrial tools. George's work integration of gas in some of his designs has earned him decent money. He has to have the necessary tools. The stoves he supplies to restaurants and fast-food kitchens need periodic repair and maintenance. His ingenuity also attracts other kinds of jobs apart from the stoves. For example, several times during my fieldwork, daladala buses were brought to the workshop needing front bumpers. George is famous for crafting these bumpers using aluminum sheets. To be precise in this process, he needs an electric blade to make clean and precise cuts, for headlight holes, for example. Using mere hand scissors would disrupt the aesthetics of the bumper looks. He also owns an electric circular saw that earns him extra money from collectors who use it to cut through refrigerator compressors. To open a refrigerator compressor, a collector rents the circular saw for approximately 500 TZS (0.22 USD), though there is always room for negotiation.

The brothers Charles and Mwanawane own a welding machine. They specialize in small household stoves. Some of their designs use metal coils at the top. The coils help cooking pots to sit steadily, and the stoves are good at maintaining heat. With continuous inventions like this, the welding machine has become popular among crafters, and the two brothers also make extra cash by renting it to others.

These new industrial tools show that the workers are not stagnated to so-called rudimentary tools. The workshop is in motion to become an industrial floor. This comes with dynamics that will influence relationships between workers. Some will start accumulating and probably monopolize power. The condition which is already happening among the master crafters.

Tools also influence how labor among crafters at Mahakama Ya Friji is organized. When I started fieldwork in 2018, the workshop was open-ended. Every master crafter had their own corner that they occupied with their apprentices. When I returned in 2021, they had changed the arrangement. Cubicles had been created for each master crafter and his apprentices. The compartments have two parts. The front part extends outside the main building and is where the work is mostly done. One of the reasons for this is that it is very hot inside during the day. The inner part of the cubicle is used to store products, materials, and tools.

This division happened because of conflicts over crafters taking tools from each other without asking. Everyone has a toolbox, and these are not locked during the day. I often observed people asking around for their tools until they found the person who took them. However, in addition, the new arrangement provides every individual with a working station. In the cubicles, each person is assigned a corner in which they can store their property and a space in the outer part for working. Inside, there is usually a toolbox for the master crafter, who shares with his apprentices. There are also lockers for everyone. Expensive property like mobile phones and radios are kept in the locker, together with work clothes or clothes to change into in the evening after work.

Crafting is often conducted while seated, unlike on industrial production lines where workers stand. In the workshops I visited, workers perform most of their labor while seated. The early stages of cookstove production, which involve measuring, cutting, and stitching, are better done while seated. Frequently, a crafter would stand to pick up tools or materials or to find some water. However, in the latter stages of cookstove crafting, activities like painting or fitting clay pots require standing. For that matter, every worker has a low-level seat in front of the cubicles, typically car seats. Because of the cushion, the car seats provide a certain degree of comfort while working, which usually stretches from 7 to 10 hours per day.

New apprentices are usually assigned to repetitive work. Once they master one tool or produce one part of the stove, they will move to another. Over time they will learn the whole process. Because Samwel was new to the job, he was placed at the *swarf*, the handmade sheet metal bending machine. "*Mashine ya kukunjia*" (the bending machine), as the workers call it. The swarf is an improvised machine. It's main function is to bend the sharp edges of aluminum sheets to make cookstoves safe for use. The original version of the machine was donated to Hassan's father's workshop from the United Kingdom in the 1970s. Hassan's father had a workshop in Tabora where Hassan and George were apprentices. When they moved to Dar es Salaam, they needed the bender. Because they had worked with it for years, they decided to improvise a new one. The bender is made from scraps of materials collected at the workshop, as illustrated in figure 6.1.

The swarf turns the edge of the sheets, so they don't cause injuries while using the cookstoves. Once Samwel has finished and the pieces are ready, they are sent back to the front, where Charles and Mwanawane assemble them into cookstoves. Charles mixes clay soil with cement and water during the final making stage, paints the cookstove's interior walls with the mixture, and attaches the ready-made clay pot. After this, the cookstoves are painted black and placed out back at the workshop to dry in the sun. This exact process is also repeated by George, his brother Feruzi, and his five apprentices.

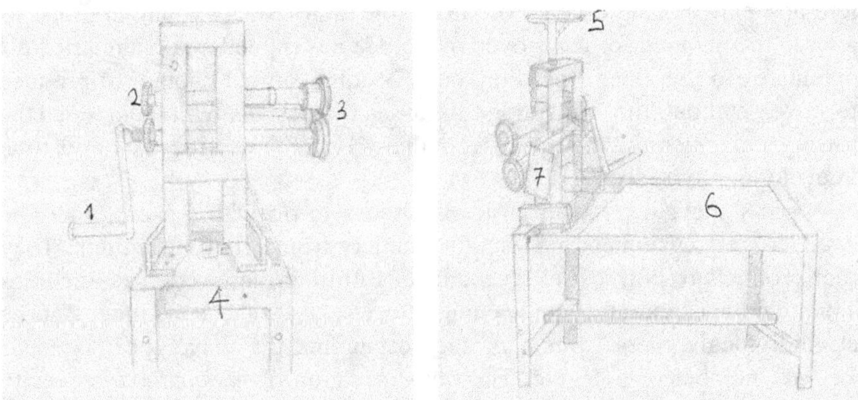

Figure 6.1 The Bender That Hassan and George Constructed from Scrap Material. Image provided by author. Note: 1 Handle made out of a bicycle pedal. The handle is attached to the lower axle. It is used to roll the axle that moves around the rail that bends the aluminum sheet. 2. Two axles scavenged from a motorcycle. The pedal is attached to the lower axle on the side with teeth. Upon pedaling, the upper axle also moves in a different direction. 3. Two sets of rails on the other end of the axle: one that looks like a train wheel and the other like a rail. The aluminum edge is pressed between the two. As the pedal rolls with the axle, the two rails bend the edge of the aluminum sheet. 4. The bender stand is made out of scrap iron. 5. The top bolt is made from a scrap iron rod. On top of it is another rod attached horizontally. The bolt is used to tighten the aluminum between the rails. 6. The bench. The plywood top is attached to the iron frame. 7. The adjuster. The metal rod is used to adjust the length of aluminum that the crafter wants to bend. Image provided by author.

PRODUCTION SPECIALIZATION

Tools also allow for the emergence of specialization. Those with simple tools, like Hassan, (though he remade the bender, which is a complicated machinery) specialize in producing a few types of objects. They cannot, for example, make the bus bumpers because doing so requires more advanced tools. George's possession of the circular saw allows him to engage with other kinds of crafting. In the meantime, the circular saw reduces his labor time compared to those using scissors to cut and shape aluminum and allows him to specialize further.

George, whom I followed the most, has developed several specializations that earn him a bit more than others at the workshop. He has developed a system that integrates gas into his BBQ stoves. Using this design, instead of using charcoal, he turns them into gas stoves needed by restaurants in the city center. The gas stoves differentiate him from the rest of the workshop. Apart from producing the stoves, his clients need him in case of faults or repairs. He is at the workshop in the mornings and afternoons. In between, he is busy traveling around the city to tend to his customers' needs. George also has more apprentices than others. The diversification attracts young crafters, as they see the demand for gas stoves rising. He has trained and maintained his apprentices to the extent that they have become skilled enough to produce the stoves without him. He simply discusses the designs with them, does the drawing and they build the stoves from there. He dedicates the rest of his time to developing hybrid stoves.

George's specialization in large BBQs has to do with the customers he gets. George's customers are usually retailers from different regions. They order products in bulk to sell in retail. In fulfilling these orders, sometimes George and his apprentice are occupied for weeks. At the same time, George gets individual orders from bars and restaurants that have specific needs like size, number of pots, etc. The transformation of his charcoal stoves to gas stoves is preferred mainly by restaurants in the city center, where gas is mainly used. The diversification of George's stoves has arisen from many years of making stoves, but also from new markets. A good example is the french-fry fryer, which he produces due to the growing fast-food and street-food culture among urban dwellers. The french-fryer is an imitation of the industrial fryers used by big restaurants. Using scrap material and new stainless steel, George imitated the fryer. For street-food businesses, George's fryer is affordable. The french-fry fryer is now his best-selling stove and has a customized size according to the needs of the customer. It is much cheaper than comparable brand-new ones.

Mwanawane and his team make only small household cookstoves, not big BBQ stoves like George. "*Tunatengeneza pekee majiko ya kina mama nyumbani*" (our stoves are for the households), he explained. The demand for household stoves is high, production costs are low, and the work is less time-consuming. In addition, they have incorporated clay pots inside their stoves to serve the charcoal. They also imitated, like George, from projects that have developed energy-saving stoves by inserting a clay pot in the burner. The combination helps the charcoal to start easier, burn more slowly, and keep the heat for longer. These three features make their stoves popular among their clients, who are predominantly women. He proudly claimed proudly that they favor their cookstoves over those from any other crafter at the workshop. It is the reason their hammers never go silent.

Hassan's return to the workshop has more meaning than missing the brotherhood and seeking social security. His formal employment with an aluminum window and door company did not allow for his ingenuity. He couldn't imagine spending his life doing that. When I returned to the workshop in 2021, Hassan took me to his cubicle. While walking there, he told me, "Sam, *njoo uone mashine mpya, unajua* one-day *tutatoboa, na huu ni mwanzo*" (Sam, come and see the new machine, you know one day we are going to make it, this is just the beginning). Hassan had invented a wind-blowing charcoal-dust stove. Usually, most charcoal dust, especially at charcoal dealers, is thrown out. There is no use for it, so he saw an opportunity to innovate a stove that would utilize the dust.

As in figure 6.2 below, the new stove is made of a car rim, which forms the spot where the charcoal dust is placed. On top of the rim, there are iron coils to hold the pot. The rim stands on three iron legs that maintain a height that allows the chef to sit. Under the rim hang two aluminum pipes, open at both ends, one facing down and the second joined to the middle of the first. The pipes keep air circulating up to the rim to keep the charcoal dust burning.

As discussed in previous chapters, the motivation to migrate to Dar es Salaam for most workers at the workshop was to search for a better future. Master crafters possessed crafting knowledge and skills before moving to the city. Their motivation was to make a better life out of those skills. The vibrant market for cookstoves and the availability of production materials incentivized their move. In Tabora, where most came from, the tobacco economy had collapsed, and materials for making watering cans for tobacco farmers had become scarce. On top of that, most households in Tabora still use traditional three-stone stoves and not charcoal stoves, compared to the big cities. With all these setbacks at home, moving to big cities was necessary.

For young members like Salum and Samwel, the motivation, of course, is the same. However, the youngsters move to the city without the skills or knowledge suitable to kickstart life in the city. Most of them are young graduates from primary and secondary schools. Their level of education is irrelevant in real life in the city, where even graduates from colleges and

Figure 6.2 Hassan Charcoal Dust Wind Stove. Photo provided by the author.

universities find it difficult to get jobs. Massive unemployment among youth in Tanzania is partly because the education system has been underfunded and understaffed and still dwells in colonial and Ujamaa-era curricula.

In the 1990s, the adoption of the structural adjustment program allowed the entry of the market into the education sector and introduced cost-sharing in social services. The cost-sharing policy became a burden to low-income families, especially women. Inequality and education gaps widened along with variations in the quality of education depending on income and area. A World Bank report of 2012 elucidates the inequalities between rural and urban, and public and private schools regarding essential infrastructure (electricity, water, and sanitation), teaching, and teachers' competence (World Bank 2012). The possibility for students from public schools to continue with higher education diminished, and the possibility of formal employment became a dream for many students from low-income areas (Matotay 2014).

At the same time, vocational training institutions started under the Small Industries Development Organization had died by the end of the 1970s. Their role and infrastructures were transferred to the Vocational Education and Training Authority (VETA), which was re-established by an Act of Parliament No. 1 of 1994[1]. However, VETA was established amid cost-sharing policies. It became challenging for many people from low-income families to access VETA training. On top of that, George and other master crafters criticize VETA for being too theoretical, and those who go to VETA have to undergo apprenticeships to master practical skills.

Living downstream does not stop people from imagining how or what the future ought to be. Lacking formal education, vocational training, or policies that safeguard their labor, discards workers' dream of a better future. Even though society labels collectors as thieves, addicts, and "crows," and the crafters as "*wagonga majiko*" (stove hammers), their imagination of a better future holds. For Hassan, Salum, and George, the phrase "one day" indicates their optimism about the future. Salum told me on one of our walks that one day when I come back to the workshop, he will own a dealership warehouse. When I met George's son, an apprentice at a car garage not far from the workshop, George jokingly said that his son becoming a good mechanic was an investment to help him in his old age. Even though his son, like his father, did not have the privilege to continue with school or go to VETA, where he could get a certificate that could land him formal employment, he imagined the future as full of possibilities and potential.

The phrase "one day" is an indeterminable, infinite time-space. "One day" is not concrete; it is something or somewhere people believe they are or should be heading, although sometimes they are not necessarily moving in that direction. Sometimes "one day" is a concrete time, a point one aims to reach. For Salum, saving for his warehouse (see chapter 5), or Hassan with his new invention, "one day" seems to be more material. George's "one day" seems more like more of a wish than something that will happen. Maybe one day, one can move from the downstream. This will probably not involve changing one's class or income.

CONCLUSION

Material consciousness alone cannot transform crafters' thoughts into material objects. George's inventions, which include gas systems and stoves made out of refrigerators, are not just intuitive creations. These works involve skills that must be developed for years, imagination, and the proper tools. The ability to diversify a product's production into many designs comes from consciousness and sense of the material a crafter works with, together with skills sharpened over the years. However, material consciousness is not enough to invent and build, collect and gather, or dismantle the scraps. Tools are central to transforming ruins, that is, electronic discards, into other objects like charcoal stoves. While it is difficult to access all necessary tools, workers improvise, reappropriate, and invent tools. The railway, which once was a tool for the empire's expansion, exploitation, and extortion and aided George and others to migrate to the city, is appropriated into the current discards politics.

Specializations that arise at the workshop, aided by different tools available, are future-oriented. Workshops like Mahakam Ya Friji, through these ingenuities, hold livelihood and socialities in many urban centers in Tanzania, let alone Dar es Salaam. George's gas stove business is growing, following the growth of urban fast-food culture and changes in access to new sources of energy. Here we see that the workshop is not on the margins but at the center of the city's fabric and changes in energy consumption. It is an integrated part that shapes the very fabric of urban life, and of urban food and the city metabolism. The temporality of the workshop and the everyday ingenuity at play among workers reflect and absorb the city itself. The workshop's interconnectedness with other social and economic relations in the city shows how discards labor and discards business are part of the city's making.

NOTE

1. The British colonial government established VETA under the Apprenticeship Ordinance of 1940. After independence, the government enacted the Vocational Training Act of 1974, establishing the National Vocational Training Division. The new VETA, established under the 1994 act, was envisioned to respond to the changing needs for skills in the labor market and to deal with burgeoning unemployment.

Chapter 7

Follow the Problem

From Discards, Charcoal Stoves, to Deforestation

Charcoal cookstoves produced at workshops like Mahakama Ya Friji are entangled with, and at the same time enhance, the sociocultural activities of preparing food and dining in the fabric of Dar es Salaam. In this way, electronic discards become more than just a digital afterlife. They are essential and integral to passionate encounters and material value. Cookstoves, social relations, and cooking are part of the broader fabric of the city, cultural performances, and environmental discussions—and all of these are connected to the activities at the Mahakama Ya Friji workshop and back to electronic discards.

I began to recognize the connection between charcoal stoves, the workshop, and the city on a weekend evening on October 28, 2018, when I decided to go out for a drink at Uhuru Peak bar. Although all this time this was under my nose, I had hardly recognized the patterns and connections. Arriving at Uhuru Peak that evening, I was welcomed by loud bongo flava music, an aroma of roasted meat, banana, and different mouth-watering soups—cow tongue, cow legs, intestines, and chicken, to name a few. As usual, when I came to Uhuru Peak bar, which got its name from the highest peak of Mount Kilimanjaro, I sat at a table near the outdoor kitchen. This table is for VIPs only. I used to come here during my master's fieldwork in 2017 and became close friends with the owners, three brothers from the Kilimanjaro region. Deo, who is always there supervising the bar, became remarkably close to me. The other two brothers were supervising a clothing shop that they also own. The youngest frequently travels to China to buy merchandise, while the eldest stays at their shop at Mwenge market.

I ordered BBQ goat ribs, *kachumbari* (guacamole) salad, and a cold Kilimanjaro beer to cool myself. Deo came around to let me know that he would join me after he finished the weekly bookkeeping. I liked coming here for several reasons. First, the bar has an outdoor area under several neem trees, suitable for my discontent with Dar es Salaam's heat. Second, I enjoyed my conversations with Deo about their (his and his brothers') business journey from artisan mining to being Machinga (street hawkers; see chapter 2) peddling shoes to now owning almost a business empire, a journey that resonates with the dreams of many workers at Mahakama Ya Friji. And, of course, I enjoyed the goat BBQ prepared meticulously by my favorite chef, Peru.

"Are you going to play some reggae?" I asked Deo after he joined me. "Ahh no," he explained as I chewed my juicy ribs, "we are playing bongo flava, you see there are many people today." I looked around to see that the place had more people than usual. Deo started telling me about the time when they first opened the bar:

> There were so many people coming here, can you imagine, on the weekends we did not have enough room to store the money. On Monday morning, we drove to the bank with money filled in backets, and those were the days. You know, all the popular people came here, musicians, politicians, army general, lawyers, everybody you care to mention was coming here. Uhuru Peak was the city's ground. You see these people, let's say there were four or five times this number. People were even standing with their drinks. Now, they come because of that legacy, and especially for the food. Peru was here with us all the time, and people still come for his BBQ.

He continued,

> *Biashara ya bar ni jiko, hatuuzi vinywaji tunauza chakula, pombe unaweza pata kila mahali. Ili watu waje kwako wanafata jiko* (In the Bar business, you do not sell the drinks, a bar is a kitchen. People come because of the food, not the alcohol we sell).

Although Uhuru Peak is no longer among the most famous bars in Dar es Salaam as it was a decade ago, now and then customers who used to come still show up. My conversations with Deo and other VIP members revealed that food from Peru draws them back to Uhuru Peak. Regular customers of Uhuru Peak include a group of lawyers who have an office nearby. Every lunchtime, they come to Uhuru Peak to eat. Although Peru admitted that the number of customers has dropped over the years, he has still managed to keep a good number of them who come regularly.

Deo's assertion that "the bar is the kitchen" can be seen in every bar in Dar es Salaam. Most bars in Dar es Salaam are constructed with a kitchen in mind. No matter how big the bar might be, the kitchen is usually filled with several charcoal stoves. Although some bars and hotels, especially in the city center and the Msasani Peninsular have slowly started switching to gas stoves, they have still not completely abandoned the famous charcoal stoves. The charcoal stove holds a vital place in assemblages of vibrant out-of-home dining in Tanzania. Apart from high-end restaurants, it is almost impossible to dine in Dar es Salaam without eating food prepared on a *jiko la mkaa* (charcoal cookstove).

Cooking and dining are social and cultural performances. The anthropology of food explores social encounters during food preparation, eating, and dining, and also explores matters of taste. Relationships surrounding food preparation, eating, and matters of taste are phenomenologically meaningful in understanding our relationships with others (people, surroundings, and objects) (Mintz 1986). Food preparation and dining experiences reveal underlying social entanglements, cultural fabrics, and material engagement. Kinship, gender relations, technologies, and relations to the more-than-humans connect to the ways we prepare and eat food.

In this chapter, I move out of the workshop to follow some of its products, the *majiko* (cookstoves), into Dar es Salaam's cityscapes and follow problems emerging together with the charcoal stoves. Although workers at Mahakama Ya Friji produce different products, including roofing materials, pans, and cooler boxes, among others, in this chapter, I have decided to focus on cookstoves because of their important role as the crafters' main product, their significance in the city's cooking and dining culture, the associated socialities in Tanzania, and the environmental problems accompanied the use of biomass for energy in Tanzania.

The cookstove assembles urban Tanzanians together in many ways: through food and sociality, in gender identities and roles or forms of masculinity and femininity, as an element of labor, and in conviviality—all the outcomes of labor at the workshops. Furthermore, the chapter explores the cookstoves' resistance in relation to and domination over new technologies and changes. I discuss how cookstoves create and contested cooking spaces and labor among city dwellers. The stove is a lens through which to witness how the lives of Tanzanians are connected by and with forms of makeshift technology.

Through this lens, workshops like Mahakama Ya Friji are at the center of food and leisure culture in Dar es Salaam. The BBQ stoves and charcoal stoves made at the workshop end up in households and dining places all around the city, solidifying the socialities around food preparation and dining. Understanding the connections between dining, cooking, and cookstoves—and how they relate back to electronic discards—contradicts a simplified categorization of workshop activities as dealing with "discards."

Unregulated electronic discards recycling, specifically in developing countries (and Tanzania in particular), needs to be understood and analyzed beyond the junkyard spaces where electronic discards are salvaged. As I have mentioned earlier, ethnography offers us the opportunity to observe processes and practices, networks, relations, and assemblages beyond the chopping, dismantling, salvaging, and burning of electronic discards. In this way, I continue gathering electronic discards as they percolate throughout wider society and follow the new or rather old problems that entangle with make-shift technologies like crafting. Ethnography allows us to follow and observe gatherings, materials, and activities, beyond recycling centers and dumping places. It allows us to follow where these materials go, what they turn into, what they gather along the way, including other environmental problems, and ways these gatherings are essential in incentivizing ways of handling electronic discards.

As I have also pointed out earlier in this book, there are studies that have explored discards' materiality and agency beyond dumping and recycling facilities. Materials salvaged from electronic discards can regain or acquire new value. However, ending there does not give a comprehensive picture of what they become. To stay with the materials means relentlessly following gatherings as they assemble along the way. In chapters 4, 5, and 6, I explored the ways in which electronic discards materials are transformed into other values inside the Mahakama Ya Friji workshop and the relations that make this possible. It is also important to follow the things beyond the confines of the workshop to explore the cookstoves and how they gather networks of labor, social relations in the city, and environmental problems.

JIKO-A KITCHEN, A STOVE, AND A WIFE

Jiko is a Swahili noun representing three different things: a place, an object, and a wife. First, jiko means a place where food is prepared, a kitchen. Second, jiko is an object, a stove that food is prepared on. And finally, jiko also means wife. There is a famous Swahili sentence, "*amepata jiko*" (he got himself a kitchen/stove), which means a man got married. From the uses of noun jiko, we can already see that the stove is supposedly a woman's object and the kitchen is a woman's place.

However, this interpretation is limited to the household sphere. Once jiko as an object or place moves outside of residences, it becomes a contested sphere. At home, jiko is a woman, a woman's object, and a woman's place. Rarely do men, especially heads of household, enter the kitchen, let alone prepare food. Men who cook at home are laughed at because they have entered the women's arena. Although men do shop for food, they rarely cook. It happens or is accepted when wives are traveling or sick. In many cases, men who happen to help their wives with kitchen work are mocked by family members and society.

Swahili has two words for the act of marrying someone, *kuoa* and *kuolewa*[1]. The first word is used only to refer to men and the second only to women. Men who cook or do kitchen chores are stripped of their masculinity and the title of head of the family, and instead are referred to as married by their wives, *wameolewa.* When a man is called *ameolewa,* it means that the woman has power over the household. There are even more derogatory ways used to describe this situation; the man might be called *mme bwege* (jerk husband), *kapigwa limbwata* (bewitched husband), *kakaliwa* (the wife is sitting on the husband), or *mwanaume suruali* (not man enough). However, a jiko outside the household is a bender. Both women and men participate in food businesses and occupy commercial kitchen spaces. However, if one looks keenly at food preparation outside the household sphere, one can see differences in the kind of food businesses that are predominately male or female. Here, I make the distinction between regular food businesses, meaning food that is usually prepared in the household, and fast-food businesses, urban street bites that usually involve deep frying. The former is dominated by women while the latter is the business of men.

The regular food includes starches like rice, *ugali*[2], and plantain banana with accompanying sauces of different kinds of fish, meat, legumes, and vegetables. Places serving these kinds of food are famously known as *Mama Lishe or Mama Ntilie*. The name itself suggests a female-dominated business. On the other hand, in fast-food and food served in recreational areas like fast-food spaces, pubs, and bars, men dominate, like Peru from Uhuru Peak bar. Fast-food in Dar es Salaam is extremely popular. Most of these places are not like fast-food outlets in the Western world (e.g., McDonalds); however, they prepare similar food, such as deep-fried french fries, fried chicken, eggs, and fish.

As I have mentioned, fast-food places are not regarded as dine out restaurants; instead, they are places to grab a bite. Many of them occupy makeshift places, perhaps just a table outside. People usually grab fries and eggs ("*chips mayai*") or chicken and chips ("*chips kuku*") when they are hungry. In contrast, eating or dining out, for the majority of people in Dar es Salaam, means *nyama choma* (BBQ grill). Nyama choma is prepared in bars and other recreational areas. Normally, people in Tanzania prefer to have nyama choma outside their homes. On rare occasions, nyama choma is grilled at home. In nyama choma places like Uhuru Peak, one can get well-prepared beef, goat, and chicken BBQ served with drinks and alcohol while Bongo Fleva or Congolese Rhumba plays in the background.

Mama Lishe usually uses a similar type of stove to those used at home—short, placed on the ground, which allows for sitting while preparing food (see figure 7.1). In contrast, in bars and fast-food places where men are predominant, they use taller stoves that allow the chef to stand while preparing

Figure 7.1 Charles Inserting Clay Pots into Charcoal Stoves. Photo provided by the author.

food. My interpretation here is that if men prepared food on the small stove while sitting, it would look the same as when women prepare food, hence the action would challenge their masculinity. The same logic applies to the kinds of foods men prepare, which are not traditionally prepared in the home kitchen. By doing this, they distance themselves from women's objects and spaces. In addition, preparing food while sitting is related to the chatting that women do when preparing food at home.

These gendered food preparation methods are reflected in the crafters' specializations as described in chapter 6. I quoted Mwanawane saying they are specialized in making cookstoves only for household users, meaning women. This is unlike George, who focused more on BBQs and stoves for restaurants and bars. George's stove designs have worked to integrate gas systems. Gas, specifically liquefied petroleum gas (LPG), is becoming cheaper, safer, and cleaner than charcoal. Restaurants, especially lavish ones, prefer gas over charcoal to prepare certain meals. Of course, charcoal BBQ stoves are still widely preferred. This is because of taste preferences, but also due to the way that BBQ has traditionally been prepared over an open fire. Again, I discussed the integration of a clay system into the stoves by Mwanawane and his kin, like those in figure 7.1. Boastfully, they spoke about how households love their stoves because of that particular integration, which saves charcoal by burning it more slowly and holding heat longer. The same kind of intervention can be seen in the unpopular energy-saving stoves.

THE SOCIALITY OF CHARCOAL STOVES

The replacement of charcoal stoves in urban centers in Tanzania is slow. The charcoal stove is dominant in every sphere of food preparation, dining, and leisure in Dar es Salaam. In Tanzania, the urban population is growing by approximately 5–6% each year. As of 2016, only 17% of the rural population and 64% of the urban population had access to electricity. Despite this, very few households use electricity for cooking due to high costs and inconsistent supply—even in urban areas. A handful of middle-class households can afford to have air conditioning or pay expensive electricity bills for cooking in the kitchen. However, even in middle-class or affluent households, a charcoal stove is always present. As a result, over 80% of the urban and peri-urban populations in Tanzania use charcoal as their primary source of energy for cooking. Across the entire country, over 90% of the 67 million inhabitants rely solely on biomass (namely charcoal and firewood) for their energy needs (Doggart et al. 2020).

This high demand for charcoal is a major cause of deforestation, forest degradation, and carbon emissions. Tanzania consumes roughly 1 million tons of charcoal each year for cooking, with half of that used in Dar es Salaam City alone (according to the World Bank in 2009). The average household in Tanzania consumes approximately 18.7 kg of charcoal per week—at a cost of about 11,000 TZS. About one hectare of forest is needed to produce roughly 3 tons of charcoal (Luoga, Witkowski, and Balkwill 2000). To address energy problems in the country, the first national energy policy was adopted in 1992, with revised policies being adopted in 2003 and 2015. All three policies have included objectives seeking to transition away from biomass energy and into electricity and fossil fuels (Doggart et al. 2020). The transition has not yet happened, and charcoal consumption keeps increasing every year.

It is interesting to note that, in Dar es Salaam, where household income has been on the rise, charcoal consumption has also increased. This goes against the "energy ladder hypothesis," which argues that households typically use biomass fuels like fuelwood and charcoal when their income is low, and then transition to "modern" fuels like kerosene, natural gas, and electricity as their income increases. The failure to transition to modern fuel is usually associated with economic reasons (Heltberg 2005; Ruiz-Mercado and Masera 2015).

While some countries have followed this transition, other countries have followed a different trajectory, in which, as households grow wealthier, they use multiple fuels in increasingly complex ways, a behavior known as "fuel stacking." Multiple reasons can account for this, including charcoal's relative price and availability, cultural preferences, and the advantages of fuel stacking over transitioning in terms of household energy efficiency and security (Makonese, Ifegbesan, and Rampedi 2018; Ruiz-Mercado and Masera 2015).

Take Aisha, my neighbor at the housing complex where I stayed during fieldwork. She is a country marketing director for an internet company. She has a good salary package that elevates her to the Tanzanian middle class. She owns a car, has moved to an apartment in an affluent neighborhood, and her son goes to an international school. Her kitchen is like any other I might find in developed countries. It is equipped with all the necessary amenities: a gas stove, a refrigerator, and a microwave. With this lifestyle, it might be assumed that she has also ascended the energy ladder. The assumption is that as people become wealthier, they move away from using biomass as a source of energy. Astonishingly, though, she still has a charcoal stove, although she only uses it for specific meals: "Rice tastes better when prepared on the charcoal stove. You know there is something with the gas stove that makes rice taste different. And by the way, I cannot cook beans or *makande* on gas."

Here, she pointed to the important issue of taste. Although I have not explored why she thinks rice tastes better on charcoal, I heard many testimonials concerning the superior taste of rice when cooked on a charcoal stove. Second, beans and makande—which is a combination of dried corn cooked together for a long time with dried beans or dried Bambara groundnut—take at least an hour and a half to cook. It is expensive to prepare these kinds of meals on gas or electricity. Although makande is not eaten frequently, beans are one of the main sources of protein and a sauce favored for rice or ugali—the stiff corn porridge—and are consumed daily by many households in Tanzania.

Here we see that middle-class households like Aisha's prefer to cook certain types of foods, such as rice and beans, on the charcoal stove, while the rest are prepared on the gas stove. This is not the case in lower-class households' kitchens, where the charcoal stove takes the primary role, occasionally supplemented by a kerosene or wood stove. Economic reasons have been cited as the driving force for why charcoal stoves are widely used in Tanzanian households (Beukering et al. 2007; Mwampamba 2007; D'Agostino, Urpelainen, and Xu 2015). There have been campaigns for over three decades to persuade households to transition from charcoal to other energy sources like electricity and gas. Even though the use of gas stoves has been on the rise among Dar es Salaam residents, the demand for charcoal stoves has not fallen, and the Mahakama Ya Friji workshop is still a vibrant place for supplying stoves and generating an impeccable livelihood for the crafters. The government, international organizations, and civil society have been at the forefront of finding more sustainable means of cooking than the use of charcoal, which has diminished forests all over the country. However, motives for transitioning have focused on economic, environmental, and health problems and ignored the social gatherings around charcoal stoves.

For a number of social reasons, Tanzanians still prefer charcoal stoves. Food preparation in low-income households is about more than just taste; it is also a social activity. It is customary in urban Tanzania to cook either in the backyard or on the front veranda. During cooking activities, women will chat with neighbors and show off what they can afford to eat. As the cooks interact with materials like pots, fire, and spices, they also interact with each other.

The materiality of the charcoal stove also matters. Most local foods have to be prepared from scratch. That means that women, who are responsible for household food preparations, have to spend long hours in the kitchen. Many Tanzanian households eat three meals a day: breakfast, usually tea and small bites ("*vitafunwa*") that vary from wheat-based donuts, bread, and chapati to rice donuts (*vitumbua*) to roots like sweet potatoes and cassava; and lunch and dinner, which usually rotate around rice, plantain, or ugali dipped in different sauces, including different types of legumes, fish, meat, and vegetables. All three meals are prepared from scratch, although breakfast is also frequently bought from Mama Lishe. Cooking is an ongoing activity in households. At least two to three hours every day are spent on preparing meals. This influences what kind of kitchen utensils should be used. It is here where the charcoal stove takes the prime position.

While most electric and gas stoves require the chef to stand, the charcoal stove is short enough to allow sitting during food preparation. It can also, by design, easily be moved around. In Dar es Salaam, the temperature is constantly above 20°C, making it nearly impossible to cook inside without an air conditioner—and most houses do not have air conditioning systems. The charcoal stove, by default, can be moved to porches or even outside under the shade of a tree. There, cooks can sit comfortably, escaping the confines of a hot kitchen. In addition, the charcoal stove is not affected by wind because charcoal does not produce high flames.

The charcoal stove has resisted the humility of things, meaning that a thing is in the background, not visible, until faults cause people to see its agency (Miller 2005). The sociality of the charcoal stove, through its position in the household, has made it visible and vital in environmental, health, and economic discussions. Charcoal stoves allow gatherings that not only foster relationships but also showcase roles, power, and status. What is cooked on top of them opens up economic interpretations and social encounters. To illustrate some of the arguments above, the following part presents three cases that span gendered, householding, and small-scale business (ujasiriamali) dynamics across the city.

Case 1: Cooking Socialities

"*Leo mama Juma akaanga nyama*" (Mama Juma is preparing beef today), my sister Mwaka tells me as we sit on her apartment's tiny back porch, facing other cooking renters. During my fieldwork, I visited my sister most weekends to chat and share dinner with her family. She lives in a two-bedroom apartment with her two teenage daughters in the Kimara area, located around 10 km from the city center.

The area hosts the new middle class of Dar es Salaam. It is normal to see four-wheel-drive cars stirring up dust as they drive to or from work in the city center. Villas with colored roofs and heavy gates can be seen climbing up the hills that make up the topography of this area. However, in the valleys, low-income families find themselves in cheaper housing. In this way, the area is occupied by different classes. Middle-class families moved here because the land was cheap, as did low-income families who moved due to cheaper rents compared to those close to the city center.

The compound my sister lives in is similar to many rental houses in Dar es Salaam. The main house is in the front, connected by rented rooms at the back to form a square shape. All doors of the rented rooms and the back door of the main house face the square in the middle, which is used for laundry, drying clothes in the sun, and cooking in front of each rented room. Each time I visited my sister, we sat on her porch chatting while she prepared dinner. All the other tenants did the same. It is hot and unbearable to cook in a confined room. I got to know some of her neighbors, who would frequently exchange words with my sister from their porches. They talked about their partners, their kids' progress in school, and developing behaviors.

They would discuss who is the best hairdresser, which shopkeeper gives a discount, or which business venture currently has more demand. They also talked about who cooks meat, rice, and other types of food and how this relates to their income. They spoke about who frequently cooks *dagaa* (dried anchovies) and *sukuma wiki* (similar to kale). This assemblage and sociality happen during cooking time, from when women start their charcoal stoves, chop and prepare food to the time the food is ready. Then everybody goes back inside for dinner.

The talks never include talk of the stoves. The conversations are about who prepared which food and an analysis of their income based on what is being cooked. At the end of the month, for those whose husbands are in formal employment, the aroma of meat or chicken comes from their stoves. At that point, the women know that that particular husband received his salary. And, for those whose husbands are employed in the informal sector, the questions will be around what deal the husband struck that has influenced any changes in the type of food prepared.

Although the stove can be seen here at the epicenter of sociality, no one ever cares to discuss each other's stoves, nor does the stove symbolize any sort of power or status. This sociality is vital because of its connection to rural-to-urban migration, as described in chapter 2. Dar es Salaam is a young city; many of its inhabitants have only recently started living in an urban environment. The majority of Dar es Salaam residents are of the first and second generations to move to urban centers from rural areas where traditional three-stone stoves still predominate. Three-stone stoves are easy to assemble. They involve three stones of the same size—which can be picked up from the environment—that hold the pot, leaving between them three pathways for firewood.

Three-stone stoves also allow the person preparing the food to sit on the ground, on a piece of wood, or on a wooden stool. This is important in rural areas where people farm crops, rear animals, or both. Women spend most of their labor time standing. Preparing food is one of those moments when women can sit and chat. The comfort of the three stones, I suggest, has been carried over into charcoal stoves, which have almost the same height as the three-stone version. As a result, the charcoal stove has not removed the sociality of cooking.

In rural areas, men, after work, will also form their social circles to chat. While waiting for dinner, they will hang out at local bars, enjoying a drink or playing board games. Here, hot local politics or football discussions will take place. Leisure time in Dar es Salaam has a similar trend. There are numerous small coffee vendors and bars in every corner of the city. In both these spaces, the charcoal stove is central to leisure activities.

Case 2: Kijiweni: Coffee, Charcoal Stove, and Socialities

I love my coffee. For that matter, a few weeks after arriving in Dar es Salaam, I had already discovered a coffee shop near my residence. It is not a formal coffee shop like Starbucks, just three benches, a table, and a charcoal stove under a neem tree next to a bus stop. In the mornings and evenings, the shop is crowded with men holding small cups filled with either thick, dark coffee or sweetened ginger tea. The old pro-ruling party generation meets here for hot political discussions with younger opposition supporters. Red-and-white Simba sports club fans joke with the yellow-and-green Dar Young Africans supporters. I spent most of my mornings and evenings here, listening and sometimes participating in the discussions.

The coffee place is referred to as *kijiwe* or *kijiweni*[3], a place where people meet and talk. People of all kinds come here. Several benches surround a wooden table under a neem tree next to Macho bus stop, one of the busiest bus stops on the edge of the Msasani Peninsula. People working in consulates, businesses, housing, cleaning, and transportation stop here every day before and after work for a cup of coffee or ginger tea.

The brewer usually opens at 5 a.m. He brings his stove and starts the fire. Within half an hour, the place gets crowded. First to come are the bus drivers and conductors who usually wake up at 4 a.m. to start their day. In the evening, the place closes after midnight as people go to bed and many buses stop shuttling. Although this is a male-dominated space, now and then women will also join while waiting for the bus. Usually, women will not talk that much compared to men. This is probably because of the topics discussed but also because it is known that vijiweni are male-dominated places.

The brewer, nicknamed Jiongeze,[4] is from the Dodoma region and is famous in Msasani. Jiongeze likes to tell stories but also to instigate conversations when people are not talking. It is a business strategy. The more that people are in intense discussion, the more coffee they drink. While cooking or cleaning cups, he will always remind those sitting around about what happened some days ago or what he has heard on the news. The discussions start from there. Sometimes, certain topics of discussion will go on for a whole day if the topic attracts enough people. One might leave, go to work, and come back to find a completely new batch of people still talking about the same topic.

Like many other young people, Jiongeze left Dodoma, the capital region, searching for a better life in Dar es Salaam. Dodoma is one of those administrative capitals with not much business or other ventures apart from government offices. Even the embassies of most countries are located in Dar es Salaam, with only a consulate in Dodoma. Dodoma is known for its semi-arid climate and rampant poverty. Much like the Nyamwezi at the workshop, the Gogo, Jiongeze's ethnic group, are famous in Dar es Salaam for their role in the coffee business.

There is no historical connection between the Gogo people and coffee. Dodoma's climate does not support coffee farming. One might assume that the Chaga people, from the slopes of Mount Kilimanjaro where coffee is grown, would be the coffee brewers. One possible explanation, similar to the one I gave about the Nyamwezi and the workshop, is that when people move from rural areas to the city, they stay with people from their area. The introduction to the labor market or jobs will heavily depend on what their hosts do. When Jiongeze arrived in Dar es Salaam, he lived with a group of

Figure 7.2 The Charcoal Stove I Built with George Is Still at Jiongeze's Coffee Shop. Photo provided by the author.

brewers from Dodoma. Years later, he hosted his younger brother, who then also opened his own coffee place.

Jiongeze sells around 50 liters of coffee and 30 liters of ginger tea every day. One cup of coffee, which is 1 deciliter, costs 50 TZS, and a 2.5 deciliter cup of ginger is also 50 TZS. The materials—charcoal, coffee, and water—cost around 6,000 TZS (3 USD). So, in a day, Jiongeze makes around 25,000 TZS (11 USD) from coffee and 6,000 TZS (3 USD) from ginger tea. Ginger tea covers his daily expenses, while the profit is earned from the coffee. Roughly, based on this daily profit, his monthly income will be around 750,000 TZS (325 USD), more than the monthly salary of a lower-level public servant.

The capital to start the kijiwe can be described as follows: one charcoal stove (the cost depends on the type; the one I built for him with double pots cost around 40,000 TZS (17 USD)); benches and tables made from recycled construction wood (I estimate that these would cost 20,000 TZS (7 USD) for four benches and a table); and cups (these cost 10,000 TZS (4 USD) per set; Jiongeze had around two sets of both coffee and ginger cups).

Jiongeze's income over the past 5 years has allowed him to purchase a herd of cattle back home in Dodoma and has provided a livelihood for his immediate family—a wife and two kids—and for his extended family. Kijiwe is not only a place for men to talk or discuss politics, but also a vibrant business. These coffee spaces are everywhere in Dar es Salaam, even in the city center where government and business offices are located. They are part of the lively cultural fabric that is supported by the charcoal stove. It's important to note that vijiwe are not permanent places—brewers open early in the morning and close late at night, which means they have to carry all of their utensils with them. The charcoal stoves, with their mobility and wind-tolerant features, are the only stoves that can be used in these places. This temporality of coffee places is shared with food vendors in the city. The charcoal stove becomes even more central in street-food shacks.

Case 3: Mama Lishe: Labor and Food Business

Around Mahakama Ya Friji, there are four food shacks owned collectively by several women. Two are located at the front of the workshop and the other two are at the back. Mama Lishe or Mama Ntilie, the names used for female food hawkers and their food shacks, typically exist in a symbiotic relationship with other labor activities. You can find a Mama Lishe shack near every construction site, car garage, or any other similar activity. This particular food business is women's labor. Although there are a considerable number of male food hawkers, still, Mama Lishe's labor, or cooking in general, is regarded as women's labor.

The food shack of Tatu and her group is located at the front of the workshop. Tatu, who is from Iringa in the southern highlands of Tanzania—like Salum, Jiongenze, and many others—left home after finishing primary school to come to Dar es Salaam, searching for a better life. Similar to other people following their kin or members of their ethnic group from rural areas to urban centers, Tatu joined her aunt in the food business before joining hands with three other young women her age to start the food shack next to the workshop. Tatu knows what kind of food everyone at the workshop likes. When everyone arrives in the morning between 7 and 9, Tatu and her crew have already been there for at least two hours. Breakfast will be ready, and she will ask everyone who arrives, "*kama kawaida*?" (the usual?). I adopted Salum's menu, a bowl of beans and two chapati, but Tatu always finish with "*chai bila sukari*" (tea without sugar). In the beginning, Tatu found it weird that I took my tea without sugar, but later chai bila sukari became the way to ask if I was having breakfast. Tatu always made sure everyone had breakfast before starting the day. "*Siku mbaya dada Tatu*"—"I am having a bad day," Dinho told Tatu one day. He meant that he did not have money for breakfast, so he would go without. Tatu walked up to him, asking, "Is that all you can say? You do not have money?" Tatu brought his breakfast and told him, "Eat first, then we worry about money."

Tatu, her team, and the other Mama Lishe are an essential part of the workshop. Without Mama Lishe's labor, it would have been hard for everyone else at the workshop and in many workplaces to perform their labor. Tatu and the company's labor gather the labor of collectors, crafters, scrap dealers, and researchers like me through the food they prepare and the charcoal stove used.

Tatu's food shack is located on the other side of the road, just a few meters from the Mahakama Ya Friji workshop. It is part of a long building running parallel to the road, with five rooms next to each other. Each room has its own large door, which means the rooms are meant for business activities. The first door, close to the workshop, is a car garage where the owner frequently buys spare parts from the gatherers, followed by a barber for men, a women's hairdressing salon, and then Tatu and her companions' food shack. Inside their room, there is a long table with plastic chairs surrounding the table. However, because of the heat, people rarely sit inside. When Tatu wasn't able to deliver food to the workshop, we would sit outside at two long tables with benches on both sides.

During lunchtime, the benches would be occupied by people who also work nearby, including officers from the District Occupational Safety and Health Authority office, which is located in front of the food shack. While eating, people engage in small talk; many times, people will come in groups and converse while having lunch. The kitchen is half covered by a corrugated iron sheet that allows the heat from the charcoal cookstoves to escape and the breeze to enter. Inside, Tatu and her companions have two cookstoves and two small stools. Usually, two cooks would be inside wearing white aprons and finishing cooking as people arrive at noon. There are always jokes directed at the cooks while they are busy serving the food.

If we focus on the relationship between Tatu and other Mama Lishe and their customers, it is clear that these groups are connected to the cookstove. Other types of stoves would be too expensive for Mama Lishe to run their businesses. These kinds of food shacks are everywhere in the city, providing food services for thousands of working people. The cookstoves that the crafters produce out of materials collected by gatherers are used by the Mama Lishe, who in turn make food for everyone.

FOLLOW THE PROBLEM: THE ENERGY CRISIS

Concerns about charcoal's contribution to environmental catastrophe in Tanzania have intensified in recent decades. Tanzania is facing an impending environmental crisis in the absence of a transition from biomass to other sustainable sources of household energy. With no viable alternatives to biomass, the focus has turned toward creating efficient and energy-saving charcoal stoves. With fewer than 30% of people in Tanzania using electricity and gas for cooking, the prospects for a transition to these sources are meager. Although there has been considerable investment in replacing traditional stoves with improved ones, people are less than enthusiastic about the new versions. A market assessment for charcoal conducted by the Tanzania Traditional Energy Development Organisation (TATEDO) claims that the dominance of the informal sector in the production of cookstoves, the localized market, people's low-income in relation to the price of improved stoves, and a lack of mechanisms to link consumer demand to the producers of improved stoves are the largest barriers (Clean Cooking Alliance 2012).

Even though economic factors hinder the adoption of energy-saving or clean energy technologies, practices and forms of sociality that take shape around cooking, as I have shown in the cases above, are often overlooked. Usually, energy-saving stoves are designed or supported by NGOs and international organizations in the name of saving women from the horrendous labor of collecting firewood or using charcoal, and associated health problems while also hoping to save forests, reduce greenhouse gases, and address climate change. However, the push toward improved stoves is solely motivated by environmental and economic reasoning. Attempts to replace charcoal cookstoves in urban Tanzania have encountered tremendous resistance among users. The last 30 years have witnessed numerous interventions in the charcoal sector, one of which has been the introduction of improved cookstoves aiming to reduce charcoal consumption. The main contributing factors for the failure of these interventions, apart from the economic one, are social.

We need to look at what it is about the charcoal cookstove that causes resistance to adopting improved ones. There are many designs of "improved" stoves. However, the designs usually focus on making the stoves sustainable. Users who are accustomed to the traditional stoves feel the difference upon switching to new improved designs. One of the common designs, which looks very similar to a traditional one, has a difference of an added clay pot to enhance charcoal burning. This change is a small but fundamental difference.

First, the clay pot increases the weight of the stove, which affects the movements that occur during cooking. As cooking is predominantly conducted outside, on verandas, or at Mama Lishe spaces, the weight hinders smooth movements by the cook depending on weather conditions such as sun, wind, or rain. Second, the clay pots tend to break over time because of the heat or when it rains while the pot is still hot. On top of that, the constant change of cooking areas in the house, as influenced by weather, rain, sun, and wind, makes the clay pot more susceptible to breaking down. These malfunctions come with two consequences. First, because it is an expensive investment, households are not often ready to replace the clay pot when it breaks. Second, because the stove can continue functioning without the clay pot, becoming more like the traditional one, users are often not too bothered about replacing the clay piece.

Things present their own unique meanings; their use is sometimes localized and entrenched within local relations and politics. Frictions between proper or improper use of objects allow us to understand the rationality of their users (Costall and Dreier 2006). Traditional charcoal stoves are regarded as improper technology because of their detrimental effects. The amount of forest lost every year, health-related effects, and the inability of the state to control the charcoal business make traditional charcoal stoves unpalatable. Efforts to replace them, however, have yielded few results.

The traditional charcoal stoves have traits that encourage resistance to new designs and technologies. The new designs seem to be undemocratic in the way they are designed. My take on improved stoves in Tanzania is that the design process has ignored the users. The design process ignored the coming together of convivialities and socialities. The main actors in the designs, be they government parastatals, NGOs and INGOs, or international financial institutions like the International Monetary Fund, tend to exclude disadvantaged households and hardly consider their agency and the materiality of stoves themselves. Designs arise from insufficient research premises and short-term surveys. Even though these efforts and designs have been implemented for over three decades, as I mentioned above, the focus has been solely on affordability and minimizing charcoal consumption. The persistence of the traditional charcoal cookstove lies in the tyranny of the object itself. It resists being replaced through its materiality and agency, and its ability to be at the center of democratic engagement, conviviality, and the creation of spaces. The way the charcoal stove brings people together around it allows for political, social, and cultural engagement.

The traditional charcoal stove is also not stagnant. The stoves' designs have undergone incremental changes. These changes have been slow and more rooted in users' needs, as well as drawing on new, improved designs. Here, the workshop becomes even more central to this discussion. If attention is paid to the ingenuities, instead of looking at the workshop as a space on the margins, solutions are already being developed. The workshops are central to addressing the more significant deforestation and climate change issues.

My walimu George, Mwanawane, and others at Mahakama Ya Friji have adopted some features of these new designs to boost their business. George's focus on producing stoves for businesses allows him to fully adopt many techniques in his designs. While recapitulating these features, Mwanawane and his companions are careful not to disturb the stoves' central functions, like the height that allows the cook to sit or the weight that permits movement of the stoves. The improved stove techniques for longer charcoal burning and heat preservation are favored by their customers. Their integration of the clay pot in their charcoal stoves is copied from the improved stoves. While they adopt these techniques, the stove must still hold the features preferred. Their clay pots, mixed a bit with cement, are much lighter and can tolerate weather patterns very well. And because they have a customer base, their customers know if or when the pot breaks, they can come back to the workshop and negotiate a better price for a new one.

Finally, there has also been an increasing focus on introducing charcoal briquettes produced from farm discards, like coconut and maize husks, in recent years. This change of focus might be because improved charcoal stoves did not become popular. Probably, the solution is not to improve the stove but rather to introduce new sources of energy. Many of the projects on alternative charcoal are in the early stages; it is hard to assess their success. However, it is worth mentioning that the significance of the charcoal stove in the Tanzanian kitchen has forced stakeholders to think of alternative sources of charcoal instead of trying to replace the stoves.

CONCLUSION: THE PLACE OF THE COOKSTOVE

In this chapter, I have discussed the importance of charcoal stoves, crafted from aluminum extracted from old refrigerators and air conditioners to produce a vital tool in Tanzanian kitchens. The complex entanglements that I have explored can only be untangled or understood when one stays with and follows gatherings. Keeping track of the object allows one to see these encounters, ruptures, and the interconnectivity that bring either a whole new world or the old fabrics that hold together sociocultural practices. The essence of a thing lies in the physical act of togetherness that produces conviviality between people (Mintz 1986). The companionability that Mintz highlights is experienced once turning into the sociality of the charcoal cookstove in households and outdoor cooking spaces.

This chapter demonstrates the ways in which the charcoal stove is an important object in the city. It gathers socialities at home, outside, and even at workplaces. It admits networks of relations and divisions of labor but also allows masculinity/femininity (good and toxic) to flourish, while at the same time managing to accommodate and challenge certain normative values. Mama Lishe and the cooking sphere in Tanzania involve intensive labor, where mostly women spend hours preparing food. But, as I have also illustrated, the Mama Lishe labor is a vital part of the workshop and labor in the city.

The case studies also showed how the cookstove is essential in politicking and political encounters. Its importance during elections, too, makes vijiwe an important space for the status quo to consolidate power and for local politics to happen. Jiongeze's kijiwe, which is located on the border between wealthy and poor neighborhoods, attracts customers from both sides. Even though there are fancy coffee shops on the peninsula, wealthy and powerful people still come to drink coffee at his kijiwe.

I have also argued that these socialities contradict accepting new things and new ways. Things present their own unique meanings; their use is sometimes localized and entrenched within local relations and politics. Frictions between proper or improper use of objects allow us to understand the rationality of their users (Costall and Dreier 2006). Traditional charcoal stoves are regarded as improper technology because of their detrimental effects. The amount of forest lost every year, health-related effects, and the inability of the state to control the charcoal business make traditional charcoal stoves unpleasant to the government, environmentalists, and (I)NGOs.

The sociality of charcoal cookstoves often outweighs the danger they pose to users' health and the environmental problems they cause through deforestation and carbon emissions. Whether operating in fast-food stalls, where people grab snacks, or in leisure places, like Uhuru Peak, the charcoal stove brings vibrant business and social gatherings. The music played at bars and nyama choma makes weekends and outings in Dar es Salaam into a proper experience of a vibrant and growing megacity. I have assembled these various pieces of evidence to show the immense importance of charcoal cookstoves in gathering socialites.

NOTES

1. Kuoa is used only for men, as they are supposed to marry a woman, and kuolewa is used for women to mean that she has been married to a man. Ameoa—singular/wameoa—plural for men getting married and ameolewa—singular/wameolewa—plural for the same act of getting married.
2. A stiff porridge dish made out of maize flour. It resembles Italian polenta but is a bit thicker and without cheese.
3. Kijiweni is singular and *vijiweni is* plural. The terms describe places where people, mainly men, spend their day discussing issues, drinking coffee, and playing games.
4. Colloquial Swahili for being innovative.

Chapter 8

Lifescaping Toxicity

Locating and Living with Pollution

Juma and Dinho, with long neem sticks, turn the burning electric cables as dense smoke permeates the air, constantly altering their position to avoid coming into direct contact with and inhaling the hazardous smoke. Their focus is on the red flames that melt down the plastic castings covering shiny copper threads. "That's money," they tell me. Fire is the representation of unregulated electronic discards recycling around the globe. Most literature on unregulated electronic discards discusses fire and how it endangers people and the environment. It is when the concept of e-waste becomes logical and sound. Toxic compounds that cause various health and environmental effects, when released from electronics, become matter out of place, and definitely no one wants to be exposed to them.

The literature on electronic waste has focused on types and concentrations of hazardous compounds and their effects on people and the environment[1]. The focus on toxicity has been on instigating change through advocacy and by confronting the power structures around toxicity and its violence (Brown 2007; Buell 1998; Carson 2002; Steingraber 2010). However, polluting companies and their ally, the state, hold power when it comes to policy enactment and changes, legality, and regulation. Cleaning up polluted environments and compensation for victims usually take a long time.

Meanwhile, a growing cadre of scholars are turning toward initiatives taken by victims of toxicity and pollution exposure (Liboiron, Tironi, and Calvillo 2018; Shapiro, Zakariya, and Roberts 2017; Tironi 2018; Tironi and Rodríguez-Giralt 2017). During the limbo between knowing about exposure and actions from the government, communities have to continue living entangled with pollution and violence (Roberts 2017). In the meantime, people must continue to live their lives in polluted environments. How do they do this? What practices do they employ to make life livable?

Initiatives that people take are usually not sourced from laboratories or scientific theories. They encompass localized knowledge, technologies, medicinal practices, traditions, and rituals. Instead of deepening the victimization of already marginalized communities, these scholars focus on "forms of life and their constituent relations, from the scale of cells to that of ways of life, are enabled, constrained and extinguished within broader power systems" (Liboiron, Tironi, and Calvillo 2018, 331). This chapter aims to contribute to this body of literature by documenting "arts of living" in polluted environments, the everyday incremental changes that communities living downstream of techno-capitalism undertake, where there is no political will nor economic prioritization to protect and safeguard the welfare of the people.

Lifescaping addresses this lack of attention to what people do when they know their exposure but have less power to mend their situation. Coined by Edelstein (2003), lifescaping comprises the changes that people attempt to make when they lose trust in the institutions, including scientific institutions, that are responsible for regulation and information. In developing countries like Tanzania, these mechanisms do not even exist. The gap in regulations and knowledge influences communities' perception of the control of their present and future. Usually, people assess their environment, act accordingly, and develop pessimistic expectations about their health (Edelstein 2003). Exposure or knowledge about a polluted environment disturbs normalcy. Exposure to hazardous materials not only changes what people do but also profoundly affects how they think about themselves, their families, and their world (Edelstein 2003, 63); hence, lifescaping is the way that life is organized in an attempt to live a normal life while being aware of exposure.

Dihno and Juma's interaction with the burning bundle of wires is the famous image of the unregulated mining of metals from electronic waste in the global south. Because the tools used are juxtaposed with tools and machines used in recycling facilities in developed countries, this image becomes real. The workers are blamed for their own hazardous exposure because of using the "wrong" tools and methods. They are also made responsible for finding their own lifescaping solutions because of a lack of scientific information and regulatory action, among other things. However, lifescaping practices—while managing exposure, as noted above—are embedded in a frame that points back to global capitalism.

BLAMING ELECTRONICS RECYCLERS

To conceptualize electronic discards, their essence can be understood as material ruins of techno-capitalist modes of production and associated contemporary consumerism. In this formulation, commodities are produced, consumed, and disposed of quickly in an exaggerated schema of endless accumulation. People frequently purchase new products to keep up with technological updates and advances, which are fertilized by planned obsolescence by the tech industry.

When electronics are discarded, they provide an avenue to think about how, where, and who harnesses or produces value out of them. Electronic discards pose a challenge of their own kind; they are embedded with precious metals that are important and needed for manufacturing. However, at the same time, they are waste and pollutants. The hazardous compounds lodged in electronics are detrimental to health and the environment. These two distinct features create friction on how best value can be exploited from e-discards without introducing hazardous chemicals to the environment.

E-discards have *phamakonic* traits, a remedy, poison, and scapegoat. The ever-increasing number of electronic devices discarded every year is astonishing. However, when we consider the amount of rare metals that are used to produce these devices makes us think again. E-discards provide solutions for depleting sources and the increasing demand for these metals. On top of that, if e-discards are mined properly, if there is a proper way, will reduce environmental and socio-economic problems (Zeng, Mathews, and Li 2018). The amount of these metals in already produced devices would significantly reduce field mining. Field mining is costly and requires more resources to operate, that is, capital, water, energy, and people. Mining uses between 7 to 10% of the energy produced globally (Grossman 2006). The ecological footprint of field mining can be reduced significantly by extracting e-discards. For example, recycling copper uses only 15 to 20% of the energy required to mine new copper ore (Grossman 2006).

The Basel Convention, which came into being to control the transboundary dumping of toxic chemicals, inclines more toward the harmful side of e-waste. The main goal of the convention is to control the dumping of hazardous materials in developing countries by developed ones. For that purpose, the Basel Convention employs a radical definition of e-waste. Basically, the Basel Convention states that anything that the owner has decided to discard is waste (Lepawsky 2018). However, the definition lacks an explanation of whether a defunct electronic device that the owner has not decided to discard can be classified as waste.

These fundamental assumptions and the origin of this definition are based on Western capitalism "worldings," where capitalists' powers also control the knowledge production and morality of other worlds (Yanagisako 2003). Ambiguity in e-discards' semantic definitions, lack of proper e-discards policics, weak implementation of the Basel and other related Conventions, and the high cost of managing, repairing, and recycling e-discards lead to massive exports of e-discards between and within regions (Lepawsky and McNabb 2010).

The analysis of e-discards resides on these two sides, just like a coin. When playing a head or tail game, the two sides are the only ones we see or choose to see. However, there is an area, very thin, between the head and tail that we do not bother to consider—the area of "Indeterminacy." Indeterminacy is the lack of recognition on someone's part, not counting the energy, skills, and ingenuity of the unregulated recycler because they do not fit the recycling politics of the day, which are capitalist worldings (Alexander and Sanchez 2018). Unregulated e-discards activities, especially in developing countries, lack recognition of their importance in the resource-hazardous chemicals categorization, so are their skills, innovation, and the value they produce. Hence, we are unable to determine future trends and probable resistances from unregulated recyclers and the materiality of discarded electronics. The agency of unregulated e-discards workers lies between the recovery and valorization of value from e-discards and exposure to hazardous chemicals.

The value-toxicity dichotomy discussed when it comes to formal and unregulated practices is negotiated, as I have described throughout this book, along the lines of the methods and tools used to extract value from electronic discards. In these negotiations, tools become political: which tools are appropriate and for which practices. As political artifacts, tools are entrenched with power, authority, and hierarchy and also determine the methods used, thereby influencing political discussions about what is proper electronic discards recycling and how it should be done. Recycling is a politically loaded concept, especially under neo-liberal economic thoughts. They determine what kinds of practices are acceptable within regulatory frameworks and which are not environmentally friendly, which are the results of the day's politics. Practices like burning, chopping, shredding, and incineration are embedded within the tools used. What is considered proper and improper during electronic discards recycling is entangled with the politics of tools and methods or practices used.

Let us pick a random e-waste recycling tool from the internet. The TOMRA Autosort Fines Optical Sorter machine from the Van Dyk Recycling Solution company's website. This is how the Autosort Fines Optical Sorter is described:

> Electronics Waste Recycling or high-volume E-scrap processing and sorting need the right tool for the job. That tool is the AUTOSORT Fines Optical Sorter! The TOMRA AUTOSORT Fines machine is the machine of choice for electronics waste recycling or E-scrap processing. It is so precise it can accurately find and sort precious metal electronic discards pieces in the 1–2 mm range. It acts as a "Swiss Army Knife" to produce a clean metals fraction, clean green circuit boards, upgrade your plastics fraction, and clean up your aluminum and other precious metals. It utilizes 4 different technologies combined in one

unit to improve your electronics recycling end products, whatever they are. (Van Dyk Recycling Solutions, n.d.)

"Precise," the "right tool," "clean," and "accurate" are some of the keywords from the company's website advertising the TOMRA Autosort Fines machine. These words are not just technical; they are political, and worlding, reflecting properness set against improper tools like those used at Mahakama Ya Friji. The classification of certain tools/machinery as right, clean, and accurate is the expansion of capitalism into spheres that have the possibility of accumulating wealth. The politics of tools make statements that help understand labor relations in the capitalist system. Tools in the capitalist system are meant to exploit humans and nonhumans for the purpose of accumulating wealth.

These politics also contribute to the worlding of electronic discards that define the tools used at Mahakama Ya Friji as the source of danger and pollution. When collectors arrive at the workshop, they have to dismantle, sort, and weigh the materials they have collected. Here, no specific tools are used; it depends on the nature of the object. Most dismantling activities use a hammer. However, sometimes there is a need to improvise. Anything can turn into a tool at this point. For example, a *tindo* (chisel) is used to penetrate objects with small openings, while for objects that are challenging to open, like refrigerator compressors, an electric circular saw is used. Compressors contain hazardous refrigerants; therefore, they are made to be difficult to open. This means the use of the saw contributes to blaming recyclers, much like their use of fire.

Then there is the standard tool found in all unregulated electronic discards recycling places: fire. Fire is preferred for burning electric cables to extract copper. Because it is a tedious job to peel one wire after another, the plastic covers' flammability makes fire a perfect tool to extract copper quickly and easily. Fire is the most vicious tool for releasing hazardous chemicals from electronic discards. Writing and showing fire is meant to make people complicit in the destruction of our planet and the creation of polluted wastelands.

TOXICITY AT THE DOWNSTREAM

Active hazardous compounds released during electronic discards salvaging is a significant concern not only for researchers and environmentalists but also for the workers themselves. As unregulated electronic discards recycling picks up in a fragile legal setting, feeding the demands of the market, the consequences of pollution and exposure to hazardous chemicals associated with electronic discards are being felt. However, because of the workshop's positionality in circulations of technology, capital, and value, the methods and tools employed in these places release the hazardous chemicals embedded in electronic discards. These chemical compounds help electronic devices function while negatively affecting the environment and health when they are not correctly handled (Sjödin 2006). A good example of these sometimes useful, sometimes nefarious substances is the Persistent Organic Pollutants (POPs).

One of the POPs are polychlorinated biphenyls (PCBs), which were phased out of use in the 1970s. PCBs were used in electrical equipment, including in cooling liquids in transformers; in dielectric fluid in large and small capacitors; as an additive in hydraulic fluids in mining equipment, vacuum pumps, laminating agents, and impregnating agents; as plasticizers in lubricants, inks, paints, adhesives, waxes, and carbonless copy paper; as casting agents and dedusting agents in cement and plaster; as well as in sealing liquids, fire retardants, immersion oils, and pesticides (Nipen et al. 2021).

In some functions, PCBs were replaced with polybrominated diphenyl ethers (PBDEs), the production of which started around the 1970s before their regulation in the 2010s. PBDEs were used as flame retardants in electronics, furniture foam, textiles, and construction materials, and vehicles (Nipen et al. 2022). However, PCBs and PBDEs are currently no longer in use because they are regulated under the Aarhus and Stockholm protocols. Some of these might still be found in electronic discards in Tanzania, given the tendency to repair items for many years before replacing them.

Other compounds include "chemicals of emerging concern" (CECs), like the alternative brominated flame retardants that replaced PBDEs in electronics to meet fire safety guidelines. The term CEC is applied to chemicals that are not yet subject to regulation, but which exhibit some or all the POP hazard criteria. There is also mercury, which at high-level concentrations was found in Dar es Salaam (Nipen et al. 2022). Although not an organic compound, it behaves in ways similar to POPs. Used in artisanal and small-scale gold mining, fossil fuel combustion, cement production, metal smelting, waste incineration, processing of electronic discards, and biomass burning, mercury is known for its persistence, accumulation in organisms, and neurotoxic properties (Nipen et al. 2021).

There are increasing concerns about the health effects of these chemical compounds embedded in electronic discards. These concerns brought about the idea that discarded electronics are waste and overshadow, especially when it comes to unregulated recycling, the resource side of it. Several reported health problems related to exposure to compounds in electronic waste include fetal loss, prematurity, low birth weight, abnormal thyroid function and thyroid development, DNA damage, and an increased risk of some chronic diseases later in life for children, such as cancer and cardiovascular disease (Grant et al. 2013; Heacock et al. 2016; WHO 2021).

Unregulated electronic discards recycling locations, like the Mahakama Ya Friji, are at the downstream end of the techno-capitalist system and are where the abovementioned hazardous compounds are dumped. Workers and the environment are exposed to these hazardous chemicals, and the sites are entangled with the "slow violence" that gradually forms networks of hazardous contamination (Davies 2019; Nixon 2011). Those living downstream experience double slow violence. First, by being at the receiving end of capital and its violence, and second, by being exposed to the toxic compounds from the materials they work with.

Although Dar es Salaam dwellers do not know much about the chemical compounds, concerns about hazardous exposures have increased in recent years. The dilemmas of exposure to hazardous chemicals and their effects are real and everyday experiences in Dar es Salaam. Urban dwellers are becoming increasingly concerned with pollution in their environment, and the quality of their health and wellness of life in general. However, scientific knowledge concerning the hazardous chemicals embedded in electronic discards is not widespread.

The dilemma has led to individual initiatives to minimize and protect themselves from exposure to hazardous chemicals. People in their everyday lives note, mark, and map possible sources of exposure and try, within the limits of their capacity, to tackle those gaps. My observations of these small-scale lifescaping initiatives suggest that there is some community awareness and lay knowledge of pollution and toxicity. Living in a polluted environment demands changes in people's lives. These hypo-interventions and attuning to the "rhythm of everydayness, set out by people to secure their conditions of existence and to render knowable otherwise intractable afflictions and matters" (Tironi 2018, 439).

LOCATING HAZARDOUS CHEMICALS

At a distance from where Juma and Dinho stood burning cables that day, George is pressing a circular saw blade as it penetrates a refrigerator compressor. Refrigerants drip, forming a small pool on the dark soil around him as two coils of copper wire emerge. On another occasion, I observed Babu Ali's, the scrap dealers, partner with an axe to cut through a car battery and pour the electrolytes, a mixture of sulfuric acid and water, onto the ground where he was standing.

The soil of Dar es Salaam, being at the coast, is sandy, white, and lacks organic materials. The dark soil around the workshop can only be explained by the burnt wires, refrigerants, car battery fluids, and engine oil layered on top of each other. The *kiangazi* (dry season) wind pushes the smoke from burning cables in all directions, and the soil receives another layer of refrigerants and electrolytes to keep in its archive. George stops cutting the refrigerator compressor lying on the dark oily soil and, holding a circular saw in his left arm, points out: "I know there is '*sumu*' [poison] in this water" [the coolant spilling out from the compressor]. George is concerned about his exposure to hazardous chemicals from electronics and understands the danger of exposure to harmful compounds through his work. "The only thing I do is to take a glass of milk every evening to clean out the poison in my body."

Exposure to hazardous chemicals is not only George's concern. There has been increased concern about hazardous chemical exposures among Tanzanians. Lacking other forms of recourse, individuals take initiatives to deal with their surroundings. Meanwhile, they hope that their government will enact policies and laws and that the international community will enter into agreements and conventions. Conventions like the Basel, Stockholm, and Bamako aim to reduce, if not prevent, transportation and dumping of hazardous substances into the environment and, ultimately, also prevent them from ending up in the bodies of vulnerable populations like George and other electronic discards recyclers in the Global South.

Toxicity is sensed by a majority of Dar es Salaam dwellers, although there are no ways for the public to establish a clear relationship between sources of hazardous compounds and the exposures they feel. However, people are aware that they live in a polluted environment. In recent years, there have been concerns about increases in noncommunicable diseases, and some people point to exposure to hazardous chemicals embedded in contemporary industrial products as one of the origins of these harms to human health. It becomes interesting to think about how sources and harms are conceived by the public.

During my participation in the activities at Mahakama Ya Friji, I noted people's concerns regarding their exposure to hazardous compounds. Despite the apparent harm from hazardous chemicals that workers are exposed to in their work, some, like George, directed their concerns toward the materials they work with, while for others, their concern was more about the hazardous chemicals in the foods they consume.

The city center and Kinondoni, where Mahakama Ya Friji is located, are separated by the Msimbazi River Valley, which makes the two places competitors in channeling rainwater into the valley. The Msimbazi River starts in the Pugu hills, 20 km southwest of the city. The river meanders through Dar es Salaam, connecting with tributaries that pull in water from all over the city. When the river reaches the ocean, it has collected detritus and waste, keeping them in its sediment core. When the rainy season starts, the activities conducted at Mahakama Ya Friji, along with other sources of hazardous compounds and waste, add another layer of hazardous chemicals to the Msimbazi Valley.

The Msimbazi catchment is well-known for urban agriculture of vegetables like *Amaranthus blitum*, *Amaranthus gangeticus*, *Cucurbita maxima*, *Ipomea batata*, and *Brassica oleracea*. The river has been declared unsafe for use, as concentrations of hazardous chemicals (notably mercury, copper, cadmium, chromium, and lead) have been found to exceed the World Health Organization's recommended levels (Leonard, Mwegoha, and Kihampa 2012; Mbuligwe and Kaseva 2005). A recent study by Maja Nipen reveals that there are increasing trends of PCBs and PBDEs in sediments from Msimbazi Valley (Nipen et al. 2022). The valley runs through congested neighborhoods with poor sanitation. There is an old landfill that continues to leach into the river, which also runs through several industrial areas that have been condemned for discharging hazardous liquids into the waterway. Dar es Salaam dwellers are fully aware of the Msimbazi River's pollution. Hence, they try to avoid vegetables and fruits that are grown near the river.

The same concerns about exposure to hazardous chemicals are felt when people try to explain the increase in cases of noncommunicable diseases like cancer and diabetes. At the time of my fieldwork, Anton, who works at Mkali's warehouse, told me that his mother had been diagnosed with cancer at the Ocean Road Cancer Institute. When I asked what he thinks of his mother's illness, he answered that cancer "*ni ugonjwa wa kisasa*" (is a modern-life disease). "Back in the day we thought cancer and diabetes were rich people's diseases, but now they are not," he added. "Even poor people get them." This is not only Anton's perception; it is shared by many in Tanzania. It is a way of understanding diseases that were not rampant before and that have few local ways of explaining the causes.

MODERNITY AND NONCOMMUNICABLE DISEASES

During my discussions at Mahakama Ya Friji and also with my friends in Tanzania, I found that people explained electronic waste as *tatizo ya kisasa* (a modern problem), the same way people would explain the growing cases of noncommunicable diseases. Electronic waste is perceived to be a result of the modern lifestyle. This modern lifestyle is also associated with certain illnesses that have become prominent in recent years, like cancer, diabetes, and hypertension. *Kisasa* (modernity) becomes a key term for understanding people's approach to toxicity. The categories of kisasa and *asili* (tradition) are juxtaposed and point to anxieties about new lifestyles as sources of harm.

Kisasa is a way of explaining new lifeways. The word "kisasa" literally translates to "of now" and is used for lifeways that cannot be explained or that do not originate in local terms. Tradition, on the other hand, relates to three terms: asili, *jadi*, and *kienyeji*. All three words mean "tradition" in English; however, in Kiswahili, they have slightly different meanings (Langwick 2011). "Asili" is "a sense of that from which things derive, an origin." Therefore, asili is used to explain the origin of a phenomenon, while "jadi" evokes "a sense of ancestry, genealogy, descent, and lineage," which also relates to another word, "*mila*" (custom). Furthermore, "kienyeji," which means "locality," relates closely to the English term "tradition" (Langwick 2011).

When kisasa is used in phrases like "*maisha ya kisasa*" (modern life), "*magonjwa ya kisasa*" (modern diseases), or "*matatizo ya kisasa*" (modern problems), it denotes a phenomenon whose origin cannot be traced, or which originates from somewhere else and has appeared recently. As kisasa originates from somewhere else, it cannot be traced back through jadi, or ancestry. Moreover, because kisasa is not of kienyeji (local) origin, it is hard to explain. These three words, "asili," "jadi," and "kienyeji," help us think about and analyze how people understand exposure to hazardous chemicals.

Although there is no direct correlation that can be drawn between new health problems in Tanzania and hazardous chemicals from electronic discards, scientific research has shown that exposure to specific hazardous electronic waste chemicals contributes to skin disease, underdevelopment of

Figure 8.1 Mama Lishe Selling Milk at Mahakama Ya Friji. Photo provided by the author.

the brain in children, damage to the nervous system, malfunctioning of the kidneys, respiratory problems, endocrine disruption, adverse pregnancy and birth outcomes, and so on (Grant et al. 2013; Heacock et al. 2016). People talk about these diseases as magonjwa ya kisasa. Although people cannot establish direct relations between exposure to hazardous chemicals and diseases, their strategies to minimize the adverse effects of these "modern problems" provide a picture of how people make sense of living in polluted environments. These acts are not specific to particular spaces, like the Mahakama Ya Friji, or to what are good recycling tools or practices. They address broader socio-political landscapes of globalized late capitalism.

MAKING SENSE OF EXPOSURE

Through my participation in the Mahakama Ya Friji, I noted a complicated relationship between people and the potential for hazardous chemical exposure. The workers varied in their concerns about and responses to (potential) toxicity. Some paid attention to potentially hazardous substances and used domestic routines or practices (like drinking milk and bathing carefully) to mitigate the effects, but others were less concerned.

One warm afternoon, I decide to withdraw from the main workshop to sit by the front deck of Mkali's scrap warehouse. I find Anton sitting on an old safe box, updating his business records. As he notices me, he closes his book and asks whether I am interested in having lunch, and he signals one of the Mama Lishe to take our orders:

> *Mkali: Kuna mboga za majani?* (Do you have vegetables?)
>
> *Mama Lishe: Ndio, kuna mchicha* (Yes, there are Amaranth leaves).
>
> *Mkali: Sio wa msimbazi huo?* (Is it not from Msimbazi Valley? [He wants to know where she got her vegetables.])
>
> *Mama Lishe: Nimenunua Buguruni* (It is from Buguruni market. [She clarifies the market she has bought the Amaranth from, which assures John that the vegetable he is ordering is not from Msimbazi.])

A portion of vegetables in a meal is crucial to many urban Tanzanians after many years of campaigns on the importance of a balanced diet that includes all the minerals and vitamins needed in the body. Lunch would usually contain ugali or rice with a selection of fish, beef curry, a bowl of beans, and a small portion of vegetables. Like Anton, many people are increasingly worried about hazardous chemicals in foods, hence his concern with locating where the vegetables came from. For him, discovering the source of his vegetables is vital.

Others, like Anton and George, were concerned with the potential health risks of hazardous exposures in their workplace. However, the overall conditions of work left them little choice in terms of risk avoidance. Instead, they practiced what Tironi (2018) calls "hypo-interventions" of a personal, somatic nature. More commonly, their concerns are directed at consumption—notably of food. Although one could not wholly avoid hazardous chemicals, consumer choices, and "lifescaping" strategies practiced within one's economic limits, allowed for some degree of risk mitigation.

One of these practices, mentioned earlier by George, is drinking milk. Tatu and other Mama Ntilie know that milk is essential to the workers at Mahakama Ya Friji as in figure 8.1. Every morning after breakfast, and in the evening before the workers start having their evening drink, the Mama Ntilie comes with thermoses full of fresh milk. They will pass around fresh milk to their clients. Every mama has her clients. It is very important for the Mama Ntilie to reserve some milk for their loyal customers. The workers form a clientele with a mama, whom they trust not to dilute the milk with water and to maintain the quality of milk. Since I didn't belong to any of the mamas, several times the mamas refused to sell milk to me, worrying that their regular clients would miss it.

Drinking milk as a detox solution is a common practice in Tanzania. Many people believe in the beneficial role of foods to "detoxify" the body (Langwick 2018). Drinking milk becomes a quotidian strategy when one is exposed to hazardous substances. As a child growing up in a rural area of Tanzania, I was made to believe that, during grazing, cattle consume different types of plants and, through digestion, they sequester and mix those plants, which will eventually become part of their milk. When a person is seriously poisoned, which in Tabora frequently happens due to snake bites, they will be given milk while being transported to the hospital. George argues that, with what he sees as a small amount or dose of sumu (poison) that he is exposed to at the workshop every day, a cup of milk is enough to clean it up.

Concerns about kisasa issues have also led to a boom in detox products from local plants that clean sumu (poison) from the body. Masai herbal detox products have become famous in Dar es Salaam, as it is believed that the group still lives *Maisha ya asili* (a traditional life). Masai herbalists are found on every corner of Dar es Salaam, and the demand for their traditional medicines has increased tremendously in recent years. Because of this demand, some herbalists like J.J. Mwaka (of For Plan International clinic) have opened traditional clinics that sell herb-based detox products and medicines and provide counseling on the danger of modern lifestyles. Mwaka and his peers advocate for a return to *vyakula vya asili* (traditional foods) as a mechanism against the modern lifestyle, which has caused health and environmental problems.

Lifescaping comprises changes like these that people make when they lose trust in the institutions responsible for regulating and providing information. When there are changes in perceptions of present and future control, people assess their environment and act accordingly, developing pessimistic expectations about health (Edelstein 2003). Hypo-interventions are the repertoire of responses of people with no power to address their exposure (Tironi 2018). These two terms allow us to bring forward the stories of people like George and others and how they act in toxic environments.

CONCLUSION

The dilemma of exposure to hazardous chemicals and their effects is an everyday experience. People make sense of their worlds in their semantic reference to kienyeji and kisasa, reluctance to eat Msimbazi vegetables, return to *vyakula vya asili* (ancestral foods), increased consumption of detox products and milk, showing different ways of explaining the life of exposure and local ways of making sense. In the bigger picture, the reluctance is also directed toward capitalism. By marking electronic waste as modern/kisasa, they also redirect the blame for hazardous chemical exposure away from their own tools and unregulated recycling practices.

Active hazardous compounds released during electronic discards salvaging are a significant concern for those exposed, although the concerns surface unevenly. As the unregulated electronic discards economy picks up in a fragile legal setting and feeds market demands, dilemmas overexposure and its effects are an everyday experience. Increasing concerns over hazardous chemical exposure among those living downstream, such as those working in unregulated electronic discards recycling, have galvanized individuals' initiatives to minimize and prevent exposure to hazardous chemicals.

In the absence of effective policy, means of advocacy, knowledge or evidence to act upon, or even weight of concern about pollution, and in a situation where pollution threats appear diffuse and hard to pinpoint, people resort to actions, choices, gestures, and practices on a microscale, hoping to protect themselves or counteract the effects of pollution. My work shows a high level of cognizant agency at work: people mark and map possible sources of exposure and try as best they can to tackle those threats. Lifescaping and hypo-interventions are done by avoiding noxious fumes, consuming "purifying" liquids like milk, or shunning vegetables from polluted areas.

All of these efforts can be seen as lifescaping initiatives and as hypo-interventions operating at a microlevel, pursued by individuals in their daily lives (Edelstein 2003; Tironi 2018). These initiatives are important because they shift our lens from pitying the "victims" of hazardous chemicals or blaming them for their own exposure to acknowledging their imaginative and agentic attempts to mitigate the known dangers of hazardous chemicals that have become deeply symbiotic with everyday lives. For the workers at the Mahakama Ya Friji, these compromises have become a typical "art of living" in polluted worlds (Tsing et al. 2017).

NOTE

1. This knowledge is too vast to cite here; Maja Nipen and Ane Haarr, from the AnthroTox project, documented it in their studies on concentrations of electronic waste-related hazardous chemicals in Dar es Salaam (Haarr et al. 2021; Nipen et al. 2021; Nipen et al. 2022). See also (Breivik et al. 2016; Grant et al. 2013; Heacock et al. 2016; Julander et al. 2014; Zeng et al. 2017).

Conclusion

Gathering Home, Hopes, and Loss

My fieldwork journey commenced in March 2018 with a reconnaissance one-month visit to Tanzania. This initial visit provided valuable insights into the electronics business, city waste management plans, and repair and recycling workshops. However, the main challenge I encountered was the inability to identify suitable sites for recycling electronic discards, leaving me perplexed. This experience underscored the complexities and challenges of fieldwork in an unregulated environment for electronic recycling and repairs. I returned to Oslo to prepare for the more extended fieldwork stay, armed with a deeper understanding of the hurdles ahead.

The pre-fieldwork in Dar es Salaam entailed visits to mobile repairers on Uhuru Street in Kariakoo. I established contacts with a few repairers, and they were ready to have me when I returned in June. However, one of the repairers mentioned another Norwegian researcher who also conducted a study with them. The problem was that the researcher never returned or wrote to them, even though they had shared their contact information. A reminder was presented on establishing and keeping relationships with interlocutors and the need to press for decolonizing fieldwork. The repairer indicated the tiredness of researchers coming to their workplaces, mingling, studying them, and leaving.

The initial plan was to conduct fieldwork in Dar es Salaam and Zanzibar. However, when I returned for the main fieldwork, I decided to do my ethnographic exploration only in Dar es Salaam. The idea was to concentrate ethnographically. Conducting ethnographic research in big cities like Dar es Salaam is complicated, it is difficult to establish meaningful relationships, and it is time-consuming. Therefore, having two field sites would have limited in-depth explorations with them.

Uhuru Street was my entry point to the field. I arrived in June and started hanging out at Uhuru Street while waiting for ethical clearance from Tanzanian authorities. I was developing rapport as an apprenticeship. At the same time, I visited other repair shops around the city and occasionally chatted with collectors on the streets. During this period, the repairers at Uhuru introduced me to more repairers outside Kariakoo. Even though I later decided to focus on recycling, the repairers on Uhuru Street were the gatekeepers, allowing for entry to other sets of actors.

After spending four months in the field, I still wondered where my field site should be: repairing, municipal waste management, or informal waste collection. As I have previously mentioned in the introduction, I went to the field with the idea that we would find massive electronic discard activities. For four months, I still needed to find where these activities were located. I spent all those months wandering around the city with mobile repairers on Uhuru Street, TV repairers in Magomeni, occasional waste collectors, and other groups. Where should I settle? Where should my focus be?

I joined the Mahakama Ya Friji workshop in October 2018, and I remained engaged with the workshop until June 2019. Unfortunately, due to the global impact of the COVID-19 outbreak, I wasn't able to resume fieldwork in 2020. However, as soon as international travel restrictions were lifted, I seized the opportunity to make a few short trips in 2021 and 2022. My most recent trip took place from September to November 2023, allowing me to continue following the work of my walimu.

Following the initial uncertainty about my study's focus, I decided to stay at the Mahakama Ya Friji in January 2020. The decision came after a fieldwork retreat with the AnthroTox team at the Amani Research Centre for Tropical Diseases in the Tanga region. I shared my confusion and frustration with senior researchers, seeking their guidance. Their advice was invaluable, guiding me on how to proceed with the fieldwork. The importance of "staying with the trouble" was emphasized, a field technique that suggested I should remain at the Mahakama Ya Friji workshop, where everything would start to fall into place. The focus on recyclers became clearer, and I began to see the fieldwork as a learning process rather than a series of challenges.

When I returned to Dar es Salaam, I decided to stay with the trouble at Mahakama Ya Friji. However, as an ethnographer conducting a study at home, it is difficult to make people trust that there is no hidden agenda. I was confident that it would be easier because I am from Tanzania and speak Swahili, which most Tanzanians speak. Even though knowing the language helped me in daily interactions and understanding the underlying meanings, I did not automatically earn their trust. It is unusual for Tanzanians to have one of their own poking and mingling with their daily business. "You do white things; I have never seen a Tanzanian researcher do what you do. You go to collect scrap. You walk all day in the sun, brother?" Salim wondered one day. I was asked, on several occasions, if I was a spy for imperialist interests. This suspicion and lack of trust were significant challenges I had to navigate in my fieldwork.

Having lived and studied in Scandinavia, I felt a sense of disconnect from the real issues happening on the ground, especially the political atmosphere of the time. Observing from afar is different from what I encountered in the field. When I started working at the workshop, everybody was skeptical about what I was doing there. They were wondering, a Tanzanian living in Norway, one of the wealthiest countries in the world, and then hanging and working with collectors, what exactly is he looking for?

One incident to point out: frequently, after a long day collecting or crafting at the workshop, where I also shared breakfast and lunch with others, I withdrew to Uhuru Peak Bar for dinner. First, the bar is near *Mahakama Ya Friji.* I could reach the place on foot after work. And second, the food is fantastic. In Dar es Salaam, the street peddlers sell their merchandise from bar to bar. I befriended one of them, who sold me pants and shoes that I had used in the field. One morning at the workshop, as I prepared to go collecting with Salum, the peddler happened to be passing by. He stopped, looking at me with a confused face. He asked why I was there and what I was doing. I replied that I was going to collect scrap with Salum. From nowhere, he started telling everybody there that he knows me; I am not the kind to collect scrap. "He cannot be collecting scraps; he is faking; I know him and his friends, *sio muokota skrepa*" (he cannot be a collector), he shouted. Salum politely explained that they knew what I was doing, and yes, I was not a scrap collector, but I went with them collecting. Salum coming to my defense was the result of months of honest discussions with workers at Mahakama Ya Friji. I explained what I was doing and translated my research proposal, questions, and ethical papers. I also presented the ethical clearance from Tanzanian authorities.

Politically, Tanzania was going through a transition in leadership, redefining the political landscape under a nationalistic approach and anti-Western rhetoric. President Magufuli's regime had created hostility and mistrust of scientific knowledge that did not favor his political interests. In July 2018, Twaweza, a nongovernmental political organization, published public survey results on the president's popularity. The president was furious about the results that suggested his popularity had dropped (AfricaNews 2018).

The incident led the Tanzanian parliament to amend the Statistics Act. The amendments infringed on scientific freedom and were set to control scientific studies that did not favor the government. Anyone who questioned the accuracy of official statistics faced a fine of at least 10 million Tanzanian shillings ($4,370), at least three years in jail, or both (Reuters 2018). The amendment came while my ethical clearance was under review. There was extra scrutiny on research projects that contributed to delaying my study.

On top of that, after the amendment of the act, government officials, NGOs, and even the public were afraid of talking to researchers. Most people were scared to provide any data that might implicate them. The situation explains why, initially, workers at Mahakama Ya Friji questioned the motive of my study. Moreover, the decision to stay at Mahakama Ya Friji and not get involved with institutions was because of the political atmosphere in Tanzania at the time.

ONE DAY: HOPES AND LOSS

Eight months since the day I saw Mjomba Mjomba crafting a cookstove, my time at Mahakama Ya Friji had come to an end. I bought five bottles of Konyagi to share with everyone, following a custom to mark events and happenings at the workshop. The sun was setting, workers had changed out of their clothes, and we gathered at the front of the workshop. The bottles were open, gin was poured, we started drinking, and cigarettes were lit. The chat was intimate, joyful, and, at some moments, sad. We reminded each other of some fun moments during my eight months at the workshop. "You are going to miss Dog Master," Dinho said, referring to the dog. I did not expect it to be hard to say goodbye.

Even though I was already thinking of returning to Oslo and starting writing, the last drink evoked emotions in everyone. I had become a member of the workshop. I had my station. "I have to get a new person for your station," George joked as he sipped his Konyagi. George had taken me on as his apprentice, giving me a station and lending me his tools. The realization of our fleeting moment of learning had come to an end. This sociality at the workshop is essential to the workers' labor.

As discussed in this book, most workshop workers were motivated to migrate to Dar es Salaam for a better future. Master crafters possessed crafting knowledge and skills before moving to the city. Their motivation was to make a better life out of those skills. Tapping into the vibrant market for cookstoves and the availability of production materials incentivized their moves. For young members like Salum and Samwel, the motivation is the same. However, the youngsters move to the city without anything to kickstart life.

However, living downstream does not stop people from imagining how or what the future ought to be—lacking formal education, vocational training, or policies that safeguard their labor, waste workers dream of a better future and take necessary action to get there. With knowledge about their exposure, life-escaping initiatives become quotidian lifeworlds. For Hassan, Salum, and George, the phrase "one day" indicated their optimism for the future, labor, and health. Salum's one day is to own a dealership warehouse. George's one day is for the son to become a good mechanic to support him when he is old. The future is imagined with full of possibilities and potential.

During my second trip to the workshop in 2021, Babu Samwel had left the workshop and moved to a village outside the city. Other workers said he thought he had become too old to work at the workshop. He retreated to the village, where he bought a plot of land for gardening and keeping a few animals. Thinking about the imaginations of the future that younger members of the workshop hold, I cannot stop wondering what Babu Samwel's imagination would have been.

The phrase "one day" is an indeterminable, infinite time-space. "One day" is not concrete; it is something or somewhere people believe they should be heading, or more so aspire to, although they are not necessarily moving in that direction. Sometimes, "one day" is a concrete temporal setting, a point one aims to reach. For Salum, saving for his warehouse, or Dullah with his new invention, "one day" seems to be more material. George's "one day" seems more like more of a wish than something that will happen. One day also connects to the city and the endless opportunities of what life tomorrow could be.

On the last trip in September 2023, I arrived at the workshop to learn that Hassan, Feruzi, Dog Master, and Simba had passed away. They passed one after the other. Hassan went first. He passed peacefully in his cubicle at the workshop during his afternoon nap. Feruzi met his fate in a work-related fatal accident while on his way to deliver charcoal stoves to a wholesale client. The Dog Master died of old age, and Simba was shot by the municipal officers following instructions from health officials after an outbreak of rabies.

Maybe one day settles in a long durée.

Bibliography

Abrahams, R. G. 1981. *The Nyamwezi Today: A Tanzanian People in the 1970's*. Changing Cultures. London: Cambridge University Press.

Adjei, Asare. 2014. "Life in Sodom and Gomorrah: The World's Largest Digital Dump via @gdnglobaldevpro." *The Guardian*. April 29, 2014. http://www.theguardian.com/global-development-professionals-network/2014/apr/29/agbogbloshie-accra-ghana-largest-ewaste-dump.

Africa Reloaded, dir. 2022. *Germany Is Helping Ghana To Build A €25 Million Electronic Waste Recycle Facility and It's Ready*. https://www.youtube.com/watch?v=jONNMh87Kmk.

AfricaNews. 2018. "Tanzania Threatens Pollster That Said Magufuli's Popularity Has Reduced by 41%." *Africanews*. Retrieved May 16, 2022. https://www.africanews.com/2018/07/13/tanzania-threatens-pollster-that-said-magufuli-s-popularity-has-reduced-by-41/.

Alber, Erdmute, and Heike Drotbohm, eds. 2015. *Anthropological Perspectives on Care: Work, Kinship, and the Life-Course*. New York: Palgrave Macmillan US. https://doi.org/10.1057/9781137513441.

Alexander, Catherine, and Andrew Sanchez. 2018. *Indeterminacy: Waste, Value, and the Imagination*. Vol. 7. New York: Berghahn Books.

Alexander, Catherine, and Joshua Reno. 2012. *Economies of Recycling: The Global Transformation of Materials, Values and Social Relations*. London, New York: Zed Books Ltd.

Andrews, Alan. 2009. "Beyond the Ban—Can the Basel Convention Adequately Safeguard the Interests of the World's Poor in the International Trade of Hazardous Waste?" *Law, Environment and Development Journal* 5 (2): 167–84.

Ayetor, G. K., Innocent Mbonigaba, M. N. Sackey, and P. Y. Andoh. 2021. "Vehicle Regulations in Africa: Impact on Used Vehicle Import and New Vehicle Sales." *Transportation Research Interdisciplinary Perspectives* 10 (June): 100384. https://doi.org/10.1016/j.trip.2021.100384.

Bagachwa, M. S. D., and A. Naho. 1995. "Estimating the Second Economy in Tanzania." *World Development* 23 (8): 1387–99. https://doi.org/10.1016/0305-750X(95)00055-H.

Bah, Mahmoud, Salmana Cisse, Bitrina Diyamett, Gouro Diallo, Fred Lerise, David Okali, Enoch Okpara, et al. 2003. "Changing Rural-Urban Linkages in Mali, Nigeria and Tanzania." *Environment & Urbanization* 15 (1): 13–23. https://doi.org/10.1630/095624703101286484.

Balasch, A., M. López, C. Reche, M. Viana, T. Moreno, and E. Eljarrat. 2022. "Exposure of Electronic Discards Dismantlers from a Formal Recycling Facility in Spain to Inhalable Organophosphate and Halogenated Flame Retardants." *Chemosphere* 294 (May): 133775. https://doi.org/10.1016/j.chemosphere.2022.133775.

Baldé, Cornelis, Vanessa Forti, Vanessa Gray, Ruediger Kuehr, and Paul Stegmann. 2017. *The Global Electronic Discards Monitor 2017: Quantities, Flows, and Resources.* Bonn; Geneva; Vienna: United Nations University (UNU), International Telecommunication Union (ITU), and International Solid Waste Association (ISWA).

BAN. 2018. "E-Trash Transparency Project." *Basel Action Network.* Retrieved November 22, 2023. https://www.ban.org/trash-transparency.

Bank, World. 2009. "Environmental Crisis or Sustainable Development Opportunity? Transforming the Charcoal Sector in Tanzania: A Policy Note." http://hdl.handle.net/10986/18603.

Barua, Maan. 2018. "Animal Work: Metabolic, Ecological, Affective." *Society for Cultural Anthropology.* https://culanth.org/fieldsights/animal-work-metabolic-ecological-affective. Accessed March 2020.

Bauman, Zygmunt. 2013. *Wasted Lives: Modernity and Its Outcasts.* Cambridge: Polity.

Beck, Ulrich. 1992. *Risk Society: Towards a New Modernity.* London; Newbury Park, CA: Sage Publications.

Bennett, Norman Robert. 1971. *Mirambo of Tanzania, 1840–1884.* London: Oxford University Press.

Beukering, P. Van, G. Kahyarara, Eric Massey, Sabina di Prima, Sebasatiaan Hess, Victor Makundi, and Kim van der Leeuw. 2007. "Optimization of the Charcoal Chain in Tanzania." https://research.vu.nl/en/publications/optimization-of-the-charcoal-chain-in-tanzania-2.

Blanchflower, David G., and Andrew J. Oswald. 1998. "What Makes an Entrepreneur." *Journal of Labor Economics* 16 (1): 35.

Boesen, Jannik, and Juhani Koponen. 1986. *Tanzania: Crisis and Struggle for Survival.* Uppsala: Nordic Africa Institute.

Breivik, Knut, James M. Armitage, Frank Wania, Andrew J. Sweetman, and Kevin C. Jones. 2016. "Tracking the Global Distribution of Persistent Organic Pollutants Accounting for Electronic discards Exports to Developing Regions." *Environmental Science & Technology* 50 (2): 798–805. https://doi.org/10.1021/acs.est.5b04226.

Brennan, James R., Andrew Burton, and Yusuf Lawi, eds. 2007. *Dar Es Salaam: Histories from an Emerging African Metropolis.* Dar es Salaam: Mkuki na Nyota Publishers.

Bromley, Ray. 1978. "Introduction—The Urban Informal Sector: Why Is It Worth Discussing?" *World Development* 6 (9): 1033–39. https://doi.org/10.1016/0305-750X(78)90061-X.

Brown, Alison. 2015. "Claiming the Streets: Property Rights and Legal Empowerment in the Urban Informal Economy." *World Development* 76: 238–48.

Brown, Phil. 2007. *Toxic Exposures: Contested Illnesses and the Environmental Health Movement*. New York: Columbia University Press.

Buell, Lawrence. 1998. "Toxic Discourse." *Critical Inquiry* 24 (3): 639–65.

Burton, Andrew. 2005. *African Underclass*. Woodbridge: James Currey Publishers.

Burton, David. 2014. *The Groundnut Line—Southern Province Railway of Tanganyika*. Telford: Shropshire.

Calas, Bernard. 2010. *From Dar Es Salaam to Bongoland: Urban Mutations in Tanzania*. Dar es Salaam: Mkuki na Nyota.

Camacho-Otero, Juana, Ida Nilstad Pettersen, and Casper Boks. 2017. "Consumer and User Acceptance in the Circular Economy: What Are Researchers Missing?" In *PLATE: Product Lifetimes and The Environment*, C. Bakker and R. Mugge (eds.), pp. 65–69. IOS Press.

Carson, Rachel. 2002. *Silent Spring*. 40th anniversary edition. Boston: Houghton Mifflin.

Chalfin, Brenda. 2019. "Waste Work and the Dialectics of Precarity in Urban Ghana: Durable Bodies and Disposable Things." *Africa: The Journal of the International African Institute* 89 (3): 499–520.

ChannelOne TV, dir. 2024. *Demolition of Agbogbloshie Market: Concerns Arise over Pace of Redevelopment Project*. https://www.youtube.com/watch?v=tcAni-8eqpNY. Accessed June 2024.

Charles, M. Judith, Dana Groskova, and Thomas M. Cahill. 2005. "Near-Source Ambient Air Monitoring of Polybrominated Diphenyl Ethers." *Report Prepared for the California Air Resources Board Project*, Project number 1–407.

Chen, Qiufan, and Ken Liu. 2019. *Waste Tide*. New York: Tor Books.

Cohen, Abner. 2014. *Urban Ethnicity*. London: Routledge.

Costall, Alan, and Ole Dreier. 2006. *Doing Things with Things: The Design and Use of Everyday Objects*. Farnham: Ashgate Publishing, Ltd.

Council of the European Union. 2024. "Circular Economy: Council Gives Final Approval to Right-to-Repair Directive." March 30, 2024. https://www.consilium.europa.eu/en/press/press-releases/2024/05/30/circular-economy-council-gives-final-approval-to-right-to-repair-directive/.

D'Agostino, Anthony L., Johannes Urpelainen, and Alice Xu. 2015. "Socio-Economic Determinants of Charcoal Expenditures in Tanzania: Evidence from Panel Data." *Energy Economics* 49 (May): 472–81. https://doi.org/10.1016/j.eneco.2015.03.007.

Dannoritzer, Cosima, dir. 2010. *The Light Bulb Conspiracy*. Documentary. Banijay Rights. https://letterboxd.com/film/the-light-bulb-conspiracy/. Accessed December 2019.

Dar es Salaam City Council. n.d. "Our Dar Es Salaam. Dar Es Salaam City Council." http://dcc.go.tz/en/our-dar-es-salaam. Accessed June 20, 2024.

Davies, Thom. 2019. "Slow Violence and Toxic Geographies: 'Out of Sight' to Whom?" *Environment and Planning C: Politics and Space*, April: 2399654419841063. https://doi.org/10.1177/2399654419841063.

De Certeau, Michel, and Pierre Mayol. 1998. *The Practice of Everyday Life: Living and Cooking*. Vol. 2. Minneapolis, MN: Universirty of Minnesota Press.

Degani, Michael. 2017. "Modal Reasoning in Dar Es Salaam's Power Network." *American Ethnologist* 44 (2): 300–14.

Denis, Jérôme, and David Pontille. 2014. "Material Ordering and the Care of Things." *Science, Technology, & Human Values* 40 (3): 38–33. https://doi.org/10.1177/0162243914553129.

DeutscheWelle. 2017. "Sakata la madawa ya kulevya Tanzania. DW. 07.02.2017." DW.COM. February 7, 2017. https://www.dw.com/sw/sakata-la-madawa-ya-kulevya-tanzania/a-37446206. Accessed January 2020.

Dirks, Nicholas B. 1992. "Castes of Mind." *Representations* 37: 56–78. https://doi.org/10.2307/2928654.

Doggart, Nike, Remidius Ruhinduka, Charles K. Meshack, Romanus C. Ishengoma, Theron Morgan-Brown, Jumanne M. Abdallah, Dominick V. Spracklen, et al. 2020. "The Influence of Energy Policy on Charcoal Consumption in Urban Households in Tanzania." *Energy for Sustainable Development* 57 (August): 200–13. https://doi.org/10.1016/j.esd.2020.06.002.

Doherty, Jacob. 2019. "Filthy Flourishing: Para-Sites, Animal Infrastructure, and the Waste Frontier in Kampala." *Current Anthropology* 60 (S20): S321–32. https://doi.org/10.1086/702868.

Doherty, Jacob. 2021. *Waste Worlds: Inhabiting Kampala's Infrastructures of Disposability*. Berkeley, CA: University of California Press.

Doherty, Jacob, and Kate Brown. 2019. "Labor Laid Waste: An Introduction to the Special Issue on Waste Work." *International Labor and Working-Class History* 95: 1–17. https://doi.org/10.1017/S0147547919000048.

Doron, Assa, and Robin Jeffrey. 2018. *Waste of a Nation: Garbage and Growth in India*. Cambridge, MA: Harvard University Press.

Douglas, Mary. 2003. *Purity and Danger: An Analysis of Concepts of Pollution and Taboo*. London: Psychology Press.

Drackner, Mikael. 2005. "What Is Waste? To Whom?—An Anthropological Perspective on Garbage." *Waste Management & Research* 23 (3): 175–81.

Dumont, Louis. 2013. "On Value: The Radcliffe-Brown Lecture in Social Anthropology, 1980." *HAU: Journal of Ethnographic Theory* 3 (1): 287–315. https://doi.org/10.14318/hau3.1.028.

Edelstein, Michael R. 2003. *Contaminated Communities: Coping with Residential Toxic Exposure*. 2nd ed. New York; London: Routledge.

Edwards, Sebastian. 2012. "Is Tanzania a Success Story." *A Long-Term Analysis*. Working Paper. National Bureau of Economic Research. https://doi.org/10.3386/w17764.

Elk, Koos van, Jan Kok, Jessica Durán, and Gert-Jan Lindeboom. 2014. "Enterprise Formalization: Fact or Fiction? A Quest for Case Studies." Deutsche Gesellschaft für Internationale Zusammenarbeit (GIZ) GmbH: International Labor

Organization. https://www.ilo.org/wcmsp5/groups/public/---ed_emp/---emp_ent/---ifp_seed/documents/publication/wcms_245359.pdf. Accessed July 2021.

Eriksen, Thomas Hylland. 2016. *Overheating: An Anthropology of Accelerated Change*. London: Pluto Press.

Esselborn, Stefan. 2013. "Environment, Memory, and the Groundnut Scheme: Britain's Largest Colonial Agricultural Development Project and Its Global Legacy." *Global Environment* 6 (11): 58–93.

E-Trash Transparency Project. n.d. Basel Action Network. https://www.ban.org/trash-transparency. Accessed November 22, 2023.

E-waste monitor. 2019. "The National E-Waste Monitor Tanzania 2019." *E-Waste Monitor*. Retrieved June 24, 2024. https://ewastemonitor.info/the-national-e-waste-monitor-tanzania-2019/.

Ewastemonitor. 2019. "Person in the Port (PiP) Study." *Electronic discards Monitor* (blog), September 13, 2019. https://ewastemonitor.info/person-in-the-port-pip-study/. Accessed December 2021.

Fahmi, Wael, and Keith Sutton. 2010. "Cairo's Contested Garbage: Sustainable Solid Waste Management and the Zabaleen's Right to the City." *Sustainability* 2 (6): 1765–83. https://doi.org/10.3390/su2061765.

Faier, Lieba, and Lisa Rofel. 2014. "Ethnographies of Encounter." *Annual Review of Anthropology* 43 (1): 363–77. https://doi.org/10.1146/annurev-anthro-102313-030210.

Forti, Vanessa, Cornelis P. Balde, Ruediger Kuehr, and Garam Bel. 2020. *The Global Electronic discards Monitor 2020: Quantities, Flows and the Circular Economy Potential*. United Nations University/United Nations Institute for Training and Research, International Telecommunication Union, and International Solid Waste Association. https://collections.unu.edu/view/UNU:7737. Accessed May 2021.

France 24. 2019. "Reporters—Inside Ghana's 'graveyard' for Europe's e-Waste." *France 24*. Retrieved June 18, 2024. https://www.france24.com/en/20190614-reporters-ghana-e-waste-electronics-toxic-pollution-agbogbloshie. Accessed February 2024.

Fredericks, Rosalind. 2018. *Garbage Citizenship: Vital Infrastructures of Labor in Dakar, Senegal*. Durham, NC: Duke University Press.

Gabrys, Jennifer. 2011. *Digital Rubbish: A Natural History of Electronics*. Ann Arbor, MI: University of Michigan Press.

George, C., and L. Olan´g. 2020. "Taxing the Informal Sector: A Case of Dar Es Salaam Street Vendors." Repoa Brief. https://www.repoa.or.tz/?publication=taxing-the-informal-sector. Accessed August 2021.

Gerasimova, Ekaterina, and Sofia Chuikina. 2009. "The Repair Society." *Russian Studies in History* 48 (1): 58–74.

Germany Embassy Accra. 2017. "Germany Supports Government of Ghana with 25 Million Euros to Tackle Environmental Impact of E-Waste." Retrieved July 1, 2024. https://accra.diplo.de/gh-en/ewaste-project-launch/1164856.

Gidwani, Vinay, and Rajyashree N. Reddy. 2011. "The Afterlives of 'Waste': Notes from India for a Minor History of Capitalist Surplus." *Antipode* 43 (5): 1625–58. https://doi.org/10.1111/j.1467-8330.2011.00902.x.

Gilligan, Carol. 1993. *In a Different Voice: Psychological Theory and Women's Development*. Cambridge: Harvard University Press.

GIZ. 2023. "Environmentally and Socially Responsible Handling of E-Waste." Retrieved July 1, 2024. https://www.giz.de/en/worldwide/123767.html.

Gordillo, Gaston. 2014. *Rubble: The Afterlife of Destruction*. Durham: Durham University Press.

Graeber, David. 2011. "'Consumption.'" *Current Anthropology* 52 (4): 489–511. https://doi.org/10.1086/660166.

Graham, Stephen, and Nigel Thrift. 2007. "Out of Order: Understanding Repair and Maintenance." *Theory, Culture & Society* 24 (3): 1–25.

Gransow, Bettina, and Zhou Daming. 2010. *Migrants and Health in Urban China*. Münster: LIT Verlag Münster.

Grant, Kristen, Fiona C. Goldizen, Peter D. Sly, Marie-Noel Brune, Maria Neira, Martin van den Berg, and Rosana E. Norman. 2013. "Health Consequences of Exposure to Electronic discards: A Systematic Review." *The Lancet Global Health* 1 (6): e350–61. https://doi.org/10.1016/S2214-109X(13)70101-3.

Gray, Hazel. 2013. "Industrial Policy and the Political Settlement in Tanzania: Aspects of Continuity and Change since Independence." *Review of African Political Economy* 40 (136): 185–201. https://doi.org/10.1080/03056244.2013.794725.

Greco, Elisa. 2015. "Landlords in the Making: Class Dynamics of the Land Grab in Mbarali, Tanzania." *Review of African Political Economy* 42 (144): 225–44. https://doi.org/10.1080/03056244.2014.992403.

Gregory, C. A. 1997. *Savage Money: The Anthropology and Politics of Commodity Exchange/C.A. Gregory*. Studies in Anthropology and History. Reading: Harwood Academic Publishers.

Gregson, Nicky. 2023. *The Waste of the World*. Bristol: Bristol University Press.

Gregson, Nicky, M. Crang, F. Ahamed, N. Akhter, and R. Ferdous. 2010. "Following Things of Rubbish Value: End-of-Life Ships, 'Chock-Chocky' Furniture and the Bangladeshi Middle-Class Consumer." *Geoforum* 41 (6): 846–54. https://doi.org/10.1016/j.geoforum.2010.05.007.

Gregson, Nicky, Alan Metcalfe, and Louise Crewe. 2009. "Practices of Object Maintenance and Repair: How Consumers Attend to Consumer Objects within the Home." *Journal of Consumer Culture* 9 (2): 248–72.

Grossman, Elizabeth. 2006. *High Tech Trash: Digital Devices, Hidden Toxics, and Human Health*. Washington: Island Press; Shearwater Books.

Haarr, Ane, Eliezer B. Mwakalapa, Aviti J. Mmochi, Jan L. Lyche, Anders Ruus, Halima Othman, Martin M. Larsen, et al. 2021. "Seasonal Rainfall Affects Occurrence of Organohalogen Contaminants in Tropical Marine Fishes and Prawns from Zanzibar, Tanzania." *Science of The Total Environment* 774 (June): 145652. https://doi.org/10.1016/j.scitotenv.2021.145652.

Hansen, Karen Tranberg. 2000. "Other People's Clothes? The International Second-Hand Clothing Trade and Dress Practices in Zambia." *Fashion Theory* 4 (3): 245–74. https://doi.org/10.2752/136270400778995499.

Haraway, Donna. 1991. *A Cyborg Manifesto: Science, Technology, and Socialist-Feminism in the Late Twentieth Century, in Simians, Cyborgs and Women: The Reinvention of Nature*. London: Routledge.

Haraway, Donna J. 2012. "Value-Added Dogs and Lively Capital." In *Lively capital: Biotechnologies, Ethics, and Governance in Global Markets*, edited by Donna J. Haraway, 93–120. Durham, NC: Duke University Press. https://doi.org/10.1215/9780822393306-003.

Harper, Douglas. 1987. *Working Knowledge: Skill and Community in a Small Shop*. Chicago: University of Chicago Press.

Harris, Marialle. 2021. "Unfinished Business: Magufuli's Autocratic Rule in Tanzania." https://www.csis.org/analysis/unfinished-business-magufulis-autocratic-rule-tanzania. Accessed March 2022.

Harriss-White, Barbara. 2020. "Waste, Social Order, and Physical Disorder in Small-Town India." *The Journal of Development Studies* 56 (2): 239–58. https://doi.org/10.1080/00220388.2019.1577386.

Hart, Keith. 1973. "Informal Income Opportunities and Urban Employment in Ghana." *The Journal of Modern African Studies* 11 (1): 61–89.

Hart, Keith. 1985. "The Informal Economy." *Cambridge Anthropology* 10 (2): 54–58. https://www.jstor.org/stable/23816368.

Havnevik, Kjell J. 1993. *Tanzania: The Limits to Development from Above*. Uppsala: Nordic Africa Institute.

Heacock, Michelle, Carol Bain Kelly, Kwadwo Ansong Asante, Linda S. Birnbaum, Åke Lennart Bergman, Marie-Noel Bruné, Irena Buka, et al. 2016. "Electronic Discards and Harm to Vulnerable Populations: A Growing Global Problem." *Environmental Health Perspectives* 124 (5): 550–55. https://doi.org/10.1289/ehp.1509699.

Heinrich, Thomas J. 1985. "Adjustment or Structural Change in Crisis Management Policy of Tanzania." *Verfassung Und Recht in Übersee/Law and Politics in Africa, Asia and Latin America* 18 (2): 195–207.

Held, Virginia. 2006. *The Ethics of Care: Personal, Political, and Global*. Oxford: Oxford University Press.

Heltberg, Rasmus. 2005. "Factors Determining Household Fuel Choice in Guatemala." *Environment and Development Economics* 10 (3): 337–61.

Henke, Christopher R. 1999. "The Mechanics of Workplace Order: Toward a Sociology of Repair." *Berkeley Journal of Sociology* 44 (1999): 55–81.

Hill, Alexandra, Tanja Hühner, Volker Kreibich, and Christian Lindner. 2014. "Dar Es Salaam, Megacity of Tomorrow: Informal Urban Expansion and the Provision of Technical Infrastructure." In *Megacities: Our Global Urban Future*, Frauke Kraas, Surinder Aggarwal, Martin Coy, and Günter Mertins (eds.), pp. 165–77. Dordrecht: Springer Netherlands. https://doi.org/10.1007/978-90-481-3417-5_12.

Hillewaert, Sarah. 2016. "'Whoever Leaves Their Traditions Is a Slave': Contemporary Notions of Servitude in an East African Town." *Africa* 86 (3): 425–46. https://doi.org/10.1017/S0001972016000322.

Hogendorn, J. S., and K. M. Scott. 1981. "The East African Groundnut Scheme: Lessons of a Large-Scale Agricultural Failure." *African Economic History* (10): 81–115. https://doi.org/10.2307/3601296.

Houston, Lara, Steven J. Jackson, Daniela K. Rosner, Syed Ishtiaque Ahmed, Meg Young, and Laewoo Kang. 2016. "Values in Repair." In *Proceedings of the 2016 CHI conference on human factors in computing systems*, by Lara Houston, Steven J. Jackson, Daniela K. Rosner, Syed Ishtiaque Ahmed, Meg Young, and Laewoo Kang, 1403–14. New York, NY: The Association for Computing Machinery.

Howell, David L. 2016. "Ethnicity and Culture in Contemporary Japan." *Journal of Contemporary History* 31 (1): 171–90. https://doi.org/10.1177/002200949603100107.

Humes, Edward. 2013. *Garbology: Our Dirty Love Affair with Trash*. London: Penguin.

Hyden, Goran. 1980. *Beyond Ujamaa in Tanzania: Underdevelopment and an Uncaptured Peasantry*. Oakland, CA: University of California Press.

Iliffe, John. 1979. *A Modern History of Tanganyika*. Oakland, CA: Cambridge University Press.

ILO. 1991. "The Dilemma of the Informal Sector." Retrieved January 26, 2022. https://ilo.primo.exlibrisgroup.com/discovery/fulldisplay/alma992806403402676/41ILO_INST:41ILO_V2.

ILO. 2002. "Decent Work and The Informal Economy: Sixth Item on the Agenda." In *International Labor Conference 90th Session*. Geneva: International Labour Office.

ILO. 2014. "Ups and Downs in the Electronics Industry: Fluctuating Production and the Use of Temporary and Other Forms of Employment." *Meeting Document*. https://www.ilo.org/resource/report/ups-and-downs-electronics-industry-fluctuating-production-and-use-temporary.

Ingold, Tim. 1997. "Eight Themes in the Anthropology of Technology." *Social Analysis: The International Journal of Social and Cultural Practice* 41 (1): 106–38.

Ingold, Tim. 2010. "Footprints through the Weather-World: Walking, Breathing, Knowing." *Journal of the Royal Anthropological Institute* 16 (s1): S121–39. https://doi.org/10.1111/j.1467-9655.2010.01613.x.

Julander, Anneli, Håkan Westberg, Magnus Engwall, and Bert Van Bavel. 2005. "Distribution of Brominated Flame Retardants in Different Dust Fractions in Air from an Electronics Recycling Facility." *Science of the Total Environment* 350 (1–3): 151–60.

Julander, Anneli, Lennart Lundgren, Lizbet Skare, Margaretha Grandér, Brita Palm, Marie Vahter, and Carola Lidén. 2014. "Formal Recycling of Electronic Discards Leads to Increased Exposure to Toxic Metals: An Occupational Exposure Study from Sweden." *Environment International* 73 (December): 243–51. https://doi.org/10.1016/j.envint.2014.07.006.

Jumbe, Aboud S., Ali Issa, and Catherine Carpenter. 2014. "Import, Handling, Disposal and Recycling of E-Waste: A Situational Analysis for Zanzibar." Unpublished.

Kelly, Robert L. 2013. *The Lifeways of Hunter-Gatherers: The Foraging Spectrum*. Cambridge: Cambridge University Press. https://doi.org/10.1017/CBO9781139176132.

Keshodkar, Akbar. 2014. "Who Needs China When You Have Dubai? The Role of Networks And The Engagement of Zanzibaris In Transnational Indian Ocean Trade." *Urban Anthropology and Studies of Cultural Systems and World Economic Development* 43 (1/2/3): 105–41.

Kiondo, Andrew S. Z. 1990. "The Second Economy in Tanzania: Its Emergence and Strategies of Control." In *The Second Economy in Marxist States*, edited by Maria Los, 175–92. Berlin: Springer.

Koponen, Juhani. 1986. *Tanzania: Crisis and Struggle for Survival*. Uppsala: Nordic Africa Institute.

Kyessi, Alphonce G., and Hussein M. Omar. 2018. "Socio-Economic Impact of Scrap Metal Business in Dar Es Alaa. The Case of Kinondoni Municipality." *Journal of Economics* 6 (4): 1–17.

Laet, Marianne de, and Annemarie Mol. 2000. "The Zimbabwe Bush Pump: Mechanics of a Fluid Technology." *Social Studies of Science* 30 (2): 225–63. https://doi.org/10.1177/030631200030002002.

Langwick, Stacey A. 2011. *Bodies, Politics, and African Healing: The Matter of Maladies in Tanzania*. Illustrated Edition. Bloomington: Indiana University Press.

Langwick, Stacey Ann. 2018. "A Politics of Habitability: Plants, Healing, and Sovereignty in a Toxic World." *Cultural Anthropology* 33 (3): 415–43. https://doi.org/10.14506/ca33.3.06.

Law, John. 1999. "After ANT: Complexity, Naming and Topology." *The Sociological Review* 47 (S1): 1–14.

Le Guin, Ursula K., Donna Jeanne Haraway, and Bul Lee. 2019. *The Carrier Bag Theory of Fiction*. London: Ignota.

Lehmacher, Wolfgang. 2017. *The Global Supply Chain : How Technology and Circular Thinking Transform Our Future*. Cham: Springer Nature.

Lepawsky, Josh. 2018. *Reassembling Rubbish: Worlding Electronic Waste*. London: The MIT Press.

Lepawsky, Josh, and Charles Mather. 2011. "From Beginnings and Endings to Boundaries and Edges: Rethinking Circulation and Exchange through Electronic Waste." *Area* 43 (3): 242–49. https://doi.org/10.1111/j.1475-4762.2011.01018.x.

Li, Tania Murray. 2010. "To Make Live or Let Die? Rural Dispossession and the Protection of Surplus Populations." *Antipode* 41 (s1): 66–93. https://doi.org/10.1111/j.1467-8330.2009.00717.x.

Liboiron, Max. 2021. *Pollution Is Colonialism*. Durham, NC: Duke University Press.

Liboiron, Max, and Josh Lepawsky. 2022. *Discard Studies: Wasting, Systems, and Power*. London: The MIT Press.

Liboiron, Max, Manuel Tironi, and Nerea Calvillo. 2018. "Toxic Politics: Acting in a Permanently Polluted World." *Social Studies of Science* 48 (3): 331–49.

Little, Peter C. 2017. "On the Micropolitics and Edges of Survival in a Technocapital Sacrifice Zone." *Capitalism Nature Socialism* 28 (4): 62–77. https://doi.org/10.1080/10455752.2016.1257037.

Little, Peter C. 2021. *Burning Matters: Life, Labor, and e-Waste Pyropolitics in Ghana*. Oxford: Oxford University Press.

Lugalla, Joe L. P. 1997. "Development, Change, and Poverty in the Informal Sector during the Era of Structural Adjustments in Tanzania." *Canadian Journal of African Studies/La Revue Canadienne Des Études Africaines* 31 (3): 424–51.

Luoga, E. J., E. T. F. Witkowski, and K. Balkwill. 2000. "Economics of Charcoal Production in Miombo Woodlands of Eastern Tanzania: Some Hidden Costs Associated with Commercialization of the Resources." *Ecological Economics* 35 (2): 243–57. https://doi.org/10.1016/S0921-8009(00)00196-8.

Luxemburg, Rosa. 2003. *The Accumulation of Capital*. London and New York: Routledge.

Maanen, John van. 1990. "Escape from Modernity: On the Ethnography of Repair and the Repair of Ethnography." Edited by Douglas Harper. *Human Studies* 13 (3): 275–84.

Magashi, A. 2011. "E-Waste Assessment Tanzania Report." *Cleaner Production*.

Makonese, Tafadzwa, Ayodeji P. Ifegbesan, and Isaac T. Rampedi. 2018. "Household Cooking Fuel Use Patterns and Determinants across Southern Africa: Evidence from the Demographic and Health Survey Data." *Energy & Environment* 29 (1): 29–48. https://doi.org/10.1177/0958305X17739475.

Malefakis, Alexis. 2019. "Creating a Market Where There Is None: Spatial Practices of Street Vendors in Dar Es Salaam, Tanzania." *City & Society* 31 (1): 119–36.

Maliyamkono, T. Luta, and Mboya S. D. Bagachwa. 1990. *The Second Economy in Tanzania*. Park Place Athens, OH: Ohio University Press.

Mans, Darius. 1994. "Tanzania: Resolute Action." In *Adjustment in Africa: Lessons from Country Case Studies*, Faruqee and Ishrat Husain (eds.), pp. 352–426. World Bank regional and sectoral studies series.

Martínez, Francisco. 2017. "Waste Is Not the End. For an Anthropology of Care, Maintenance and Repair." *Social Anthropology. John Wiley & Sons*, 25 (3): 346–50. https://doi.org/10.1111/1469-8676.12436.

Mashenene, Robert Galan. 2019. "Entrepreneurial Capabilities between the Chagga and Sukuma Owned Small and Medium Enterprises in Tanzania." Undefined. https://www.semanticscholar.org/paper/Entrepreneurial-Capabilities-between-the-Chagga-and-Mashenene/cef2142aff4f78b26d2f5288a0e0336e7f71640f.

Mavhunga, Clapperton Chakanetsa. 2014. *Transient Workspaces: Technologies of Everyday Innovation in Zimbabwe*. London: The MIT Press.

Mavhunga, Clapperton Chakanetsa, ed. 2017. *What Do Science, Technology, and Innovation Mean from Africa?* London: The MIT Press.

Medaas, Christian. 2024. "Between Technology and Sustainability: An Ethnography of Smart Phone Repair in Dar es Salaam, Tanzania and Oslo, Norway." Thesis (Unpublished).

Millar, Kathleen. 2008. "Making Trash into Treasure: Struggles for Autonomy on a Brazilian Garbage Dump." *Anthropology of Work Review* 29 (2): 25–34. https://doi.org/10.1111/j.1548-1417.2008.00011.x.

Millar, Kathleen M. 2014. "The Precarious Present: Wageless Labor and Disrupted Life in Rio de Janeiro, Brazil." *Cultural Anthropology* 29 (1): 32–53. https://doi.org/10.14506/ca29.1.04.

Millar, Kathleen M. 2018. *Reclaiming the Discarded: Life and Labor on Rio's Garbage Dump*. Durham, NC: Duke University Press.

Miller, Daniel. 2005. *Materiality*. Durham, NC: Duke University Press.

Miller, Daniel. 2012. *Consumption and Its Consequences*. Cambridge: Polity Press.

Mintz, Sidney. 1986. *Sweetness and Power. The Place of Sugar in Modern History*. 1st Edition. London: Penguin Books.

Mjema, G. D. 1999. "Youth Unemployment in Tanzania, Nature Magnitude and Proposals for Possible Solution." Dar es Salaam: ILO and The Ministry of Labor and Youth Department: Economic and Social Research Foundation.

Mol, Annemarie, Ingunn Moser, and Jeannette Pols. 2015. *Care in Practice: On Tinkering in Clinics, Homes and Farms*. Bielefeld: transcript Verlag.

Moore, Jason W. 2016. *Anthropocene or Capitalocene? Nature, History, and the Crisis of Capitalism*. Binghamton, NY: Pm Press.

Msoka, Colman Titus. 2005. *Informal Markets and Urban Development: A Study of Street Vending in Dar Es Salaam, Tanzania*. Minneapolis, MN: University of Minnesota.

Murthy, Venkatesha, and Seeram Ramakrishna. 2022. "A Review on Global Electronic discards Management: Urban Mining towards a Sustainable Future and Circular Economy." *Sustainability* 14 (2): 647. https://doi.org/10.3390/su14020647.

Mwampamba, Tuyeni Heita. 2007. "Has the Woodfuel Crisis Returned? Urban Charcoal Consumption in Tanzania and Its Implications to Present and Future Forest Availability." *Energy Policy* 35 (8): 4221–34. https://doi.org/10.1016/j.enpol.2007.02.010.

Myers, Garth Andrew. 2011. *African Cities: Alternative Visions of Urban Theory and Practice*. London: Zed Books.

Natura Urbana, dir. 2020. *Natura Urbana: The Brachen of Berlin*. Documentary. Vereinigtes Königreich. https://www.youtube.com/watch?v=Ufs81sDxtw0&ab_channel=NaturaUrbana.

Neary, Ian J. 1986. "Socialist and Communist Party Attitudes towards Discrimination against Japan's Burakumin." *Political Studies* 34 (4): 556–74. https://doi.org/10.1111/j.1467-9248.1986.tb01613.x.

Nellist, Roger. 2017. "The Tanzania Mineral Saga." September 1, 2017. https://www.tzaffairs.org/2017/09/.

Nelson, Fred, Emmanuel Sulle, and Edward Lekaita. 2012. "Land Grabbing and Political Transformation in Tanzania." In *International Conference on Global Land Grabbing II*, 17–19.

Nguyen, Minh T. N. 2018. *Waste and Wealth: An Ethnography of Labor, Value, and Morality in a Vietnamese Recycling Economy*. 1st edition. New York: Oxford University Press.

Nipen, Maja, Rolf David Vogt, Pernilla Bohlin-Nizzetto, Katrine Borgå, Eliezer Brown Mwakalapa, Anders Røsrud Borgen, Martin Schlabach, et al. 2022. "Increasing Trends of Legacy and Emerging Organic Contaminants in a Dated Sediment Core From East-Africa." *Frontiers in Environmental Science*, 9–711.

Nipen, Maja, Rolf Vogt, Pernilla Bohlin-Nizzetto, Katrine Borgå, Eliezer Mwakalapa, Anders Borgen, Susanne Jørgensen, et al. 2021. "Spatial Trends of Chlorinated Paraffins and Dechloranes in Air and Soil in a Tropical Urban, Suburban, and Rural Environment." *Environmental Pollution* 292 (October): 118298. https://doi.org/10.1016/j.envpol.2021.118298.

Nixon, Rob. 2011. *Slow Violence and the Environmentalism of the Poor*. Cambridge, MA: Harvard University Press.

NRK, dir. 2020. *The Waste Smugglers*. https://tv.nrk.no/serie/brennpunkt/ekstramateriale/MDDP11001519.

Ntapanta, Samwel Moses. 2021. "'Lifescaping' Toxicants: Locating and Living with electronic discards in Tanzania." *Anthropology Today* 37 (4): 7–10. https://doi.org/10.1111/1467-8322.12663.

Ntapanta, Samwel Moses. 2023. "Polarized Cityscapes: Gathering Electronic Waste and Its Malcontents in Dar Es Salaam." *Norsk Antropologisk Tidsskrift* 33 (3–4): 227–43. https://doi.org/10.18261/nat.33.3-4.4.

Ntapanta, Samwel Moses, and Christian Medaas. 2024. "Who Cares, and for What? Exploring the Contents and Extents of Care as an Analytical Concept in Ethnographies of Electronics Repair." *Human Organization*. (Manuscript)

O'Neill, Kate. 2019. *Waste*. Hoboken, NJ: John Wiley & Sons.

Orr, Julian E. 2016. *Talking about Machines*. Ithaca, NY: Cornell University Press.

Packard, Vance. 1969. *The Waste Makers*. New York: D. McKay Co.

Parkin, David J. 1969. "Tribe as Fact and Fiction in an East African City." In *Tradition and Transition in East Africa*, edited by in P. H. Gulliver, 273–96. Berkeley: University of California Press.

Parry, Jonathan P. 2003. "Nehru's Dream and the Village 'Waiting Room': Long-Distance Labor Migrants to a Central Indian Steel Town." *Contributions to Indian Sociology* 37 (1–2): 217–49. https://doi.org/10.1177/006996670303700110.

Pettersson-Julander, Anneli, Bert van Bavel, Magnus Engwall, and Håkan Westberg. 2004. "Personal Air Sampling and Analysis of Polybrominated Diphenyl Ethers and Other Bromine Containing Compounds at an Electronic Recycling Facility in Sweden." *Journal of Environmental Monitoring* 6 (11): 874–80.

Plessis, Pierre du. 2018. "Gathering the Kalahari: Tracking Landscapes in Motion." UC Santa Cruz. Thesis.

Plessis, Pierre du. 2022. "Tracking Meat of the Sand: Noticing Multispecies Landscapes in the Kalahari." *Environmental Humanities* 14 (1): 49–70. https://doi.org/10.1215/22011919-9481429.

Plessis, Pierre L. du. 2023. "Hunting and Gathering Practices of the Anthropocene: In Conversation about Hunting-Gathering." *Etnofoor* 35 (1): 127–32.

Polanyi, Karl. 1944. *The Great Transformation*. New York: Farrar & Rinehart, Inc.

Portes, Alejandro, and William Haller. 1985. "The Informal Economy." *Cambridge Anthropology* 10 (2): 54–58.

Purdon, Mark. 2013. "Land Acquisitions in Tanzania: Strong Sustainability, Weak Sustainability and the Importance of Comparative Methods." *Journal of Agricultural and Environmental Ethics* 6 (26): 1127–56.

Puthussery, Honeylal. 2019. "Ways of Being of Equipment: A Heideggerian Enquiry into Design Process." *Tattva Journal of Philosophy* 11 (1): 31–52.

Raja, Masood. 2019. "Worlding." *Postcolonial Space* (blog). May 2, 2019. Retrieved January 2024. https://postcolonial.net/glossary/worlding/.

Rathje, William L., and Cullen Murphy. 2001. *Rubbish!: The Archaeology of Garbage*. Tucson, AZ: University of Arizona Press.

Reno, Joshua. 2009. "Your Trash Is Someone's Treasure: The Politics of Value at a Michigan Landfill." *Journal of Material Culture* 14 (1): 29–46. https://doi.org/10.1177/1359183508100007.

Reno, Joshua. 2016. *Waste Away: Working and Living with a North American Landfill*. Oakland, CA: University of California Press.

Reuters. 2018. "Tanzania Law Punishing Critics of Statistics 'Deeply Concerning': World Bank." *Reuters*. October 3. Retrieved July 2019.

Roberts, Elizabeth F. S. 2017. "What Gets Inside: Violent Entanglements and Toxic Boundaries in Mexico City." *Cultural Anthropology* 32 (4): 592–619. https://doi.org/10.14506/ca32.4.07.

Rockel, Stephen J. 2000. "'A Nation of Porters': The Nyamwezi and the Labor Market in Nineteenth-Century Tanzania." *The Journal of African History* 41 (2): 173–95.

Rogers, Heather. 2006. *Gone Tomorrow: The Hidden Life of Garbage*. New York, NY: The New Press.

Roy, Arundhati. 2008. *The God of Small Things*. New York, NY: Random House Publishing Group.

Rufus, Anneli, and Kristan Lawson. 2009. *The Scavengers' Manifesto*. New Yrok; London: Jeremy P. Tarcher; Penguin.

Ruiz-Mercado, Ilse, and Omar Masera. 2015. "Patterns of Stove Use in the Context of Fuel-Device Stacking: Rationale and Implications." *EcoHealth* 12 (1): 42–56. https://doi.org/10.1007/s10393-015-1009-4.

Ryan, Alan. 1988. *Property*. Minnesota: University of Minnesota Press.

Samel, Swapna. 2009. "Burakumin—A Japanese Marginal Group: Japan's Hidden People Fight to Gain Equality." *Proceedings of the Indian History Congress* 70: 785–94. https://www.jstor.org/stable/44147725.

Scott, James C. 2008. *Seeing like a State*. New Haven, CT: Yale university Press.

Sennett, Richard. 2009. *The Craftsman*. New Haven, CT: Yale University Press.

Shapiro, Nicholas, Nasser Zakariya, and Jody Roberts. 2017. *A Wary Alliance: From Enumerating the Environment to Inviting Apprehension*. Oakland, CA: University of California.

Sherrington, Richard. 2007. "Mob Justice, Metaphysical Punishment and the Moralisation of Accumulation in Urban Tanzania." *Cambridge Anthropology* 27 (1): 1–24. https://www.jstor.org/stable/23820707.

Shivji, Issa G., Saida Yahya-Othman, and Kamata Ng′wanza. 2020. *Development as Rebellion: A Biography of Julius Nyerere*. Dar es salaam: Mkuki na Nyota Publishers.

Simone, AbdouMaliq. 2004. "People as Infrastructure: Intersecting Fragments in Johannesburg." *Public Culture* 16 (3): 407–29.

Simone, AbdouMaliq. 2022. *The Surrounds: Urban Life within and beyond Capture*. Durham, NC: Duke University Press. https://doi.org/10.1215/9781478022749.

Sjödin, Andreas. 2006. "High Tech Trash: Digital Devices, Hidden Toxins, and Human Health." *Environmental Health Perspectives* 114 (8): A500.

Skarstein, Rune. 2005. "Economic Liberalization and Smallholder Productivity in Tanzania. From Promised Success to Real Failure, 1985–1998." *Journal of Agrarian Change* 5 (3): 334–62. https://doi.org/10.1111/j.1471-0366.2005.00104.x.

Smiley, Sarah L. 2013. "Researching Housing, Water, and Sanitation in the British and Tanzania National Archives." *History in Africa* 40: 353–64. https://www.jstor.org/stable/26362076.

Sormani, Philippe, Alain Bovet, and Ignaz Strebel. 2019. "Introduction: When Things Break Down." In *Repair Work Ethnographies*, Ignaz Strebel, Alain Bovet, Philippe Sormani (eds.), pp. 1–27. Berlin: Springer.

Spivak, Gayatri Chakravorty. 1985. "The Rani of Sirmur: An Essay in Reading the Archives." *History and Theory* 24 (3): 247–72.

Statista. 2021. "Global E-Waste Generation Outlook 2030." *Statista*. Retrieved September 3, 2021. https://www.statista.com/statistics/1067081/generation-electronic-waste-globally-forecast/.

Steiler, Ilona. 2018. "What's in a Word? The Conceptual Politics of 'Informal' Street Trade in Dar Es Salaam." *Articulo—Journal of Urban Research* 17–18 (January). https://doi.org/10.4000/articulo.3376.

Steiler, Ilona, and Chediel Nyirenda. 2021. "Towards Sustainable Livelihood in the Tanzanian Informal Economy: Facilitating Inclusion, Organization, and Rights for Street Vendors." wp-2021-53. *WIDER Working Paper Series*. WIDER Working Paper Series. World Institute for Development Economic Research (UNU-WIDER). https://ideas.repec.org/p/unu/wpaper/wp-2021-53.html.

Stein, Howard. 1985. "Theories of the State in Tanzania: A Critical Assessment." *The Journal of Modern African Studies* 23 (1): 105–23.

Steingraber, S. 2010. *Living Downstream: An Ecologist's Personal Investigation of Cancer and the Environment*. New York, NY: Hachette Books.

Stoler, Ann Laura. 2008. "Imperial Debris: Reflections on Ruins and Ruination." *Cultural Anthropology* 23 (2): 191–219. https://doi.org/10.1111/j.1548-1360.2008.00007.x.

Strasser, Susan. 2000. *Waste and Want: A Social History of Trash*. New York, NY: Henry Holt and Company.

Strebel, Ignaz, Alain Bovet, and Philippe Sormani. 2018. *Repair Work Ethnographies: Revisiting Breakdown, Relocating Materiality*. Berlin: Springer.

Sulle, Emmanuel, and Fred Nelson. 2009. *Biofuels, Land Access and Rural Livelihoods in Tanzania*. London: IIED.

The Citizen. 2018. "JPM Leaves Petty Traders Smiling." *The Citizen*. December 10, 2018. https://www.thecitizen.co.tz/tanzania/news/jpm-leaves-petty-traders-smiling--2664362. Retrieved July 2019.

The Global E-waste Monitor. 2024. "The Global E-Waste Monitor 2024." *E-Waste Monitor*. Retrieved June 24, 2024. https://ewastemonitor.info/the-global-e-waste-monitor-2024/. Retrieved June 2024.

Thompson, Michael. 2017. *Rubbish Theory: The Creation and Destruction of Value*. 2nd Edition. London: Pluto Press.

Tironi, Manuel. 2018. “Hypo-Interventions: Intimate Activism in Toxic Environments.” *Social Studies of Science* 48 (3): 438–55. https://doi.org/10.1177/0306312718784779.

Tironi, Manuel, and Israel Rodríguez-Giralt. 2017. “Healing, Knowing, Enduring: Care and Politics in Damaged Worlds.” *The Sociological Review* 65 (2_suppl): 89–109. https://doi.org/10.1177/0081176917712874.

Tripp, Aili Mari. 1990. “The Urban Informal Economy and the State in Tanzania.” Doctoral dissertation. Illinois: Northwestern University.

Tsing, Anna Lowenhaupt. 2015a. “Salvage Accumulation, or the Structural Effects of Capitalist Generativity.” *Society for Cultural Anthropology*. https://culanth.org/fieldsights/salvage-accumulation-or-the-structural-effects-of-capitalist-generativity.

Tsing, Anna Lowenhaupt. 2015b. *The Mushroom at the End of the World: On the Possibility of Life in Capitalist Ruins*. Princeton: Princeton University Press.

Tsing, Anna Lowenhaupt, Heather Anne Swanson, Elaine Gan, and Nils Bubandt. 2017. *Arts of Living on a Damaged Planet*. Minneapolis, MI: University of Minnesota Press.

Turner, Bertram. 2017. “The Anthropology of Property.” In *Comparative Property Law*, Graziadei, Michele and Smith, Lionel (eds.), pp. 26–48. Edward Elgar Publishing.

Uden, Lorna, and Janet Francis. 2011. “Service Innovation Using Actor Network Theory.” In *Actor-Network Theory and Technology Innovation: Advancements and New Concepts*, Arthur Tatnall (ed.), pp. 20–40. Hershey, PA: IGI Global.

Unomah, A. C. 1977. *Mirambo of Tanzania*. Vol. 12. African Historical Biographies. London: Heinemann.

Veblen, Thorstein, and Robert Lekachman. 1994. *The Theory of the Leisure Class*. Penguin Twentieth Century Classics. New York: Penguin Books.

Wang, X., and G. Gaustad. 2012. “Prioritizing Material Recovery for End-of-Life Printed Circuit Boards.” *Waste Manage* 32 (10): 1903.

Weber, Max. 1958. “Science as a Vocation.” *Daedalus* 87 (1): 111–34.

Weigensamer, F., and C. Krönes, dirs. 2018. *Welcome To Sodom*. Documentary. Blackbox FilmGreenpeaceLand Steiermark, department 9—Culture, Europe, Ext. Relations.

Wendling, Amy. 2009. *Karl Marx on Technology and Alienation*. Berlin: Springer.

WHO. 2021. “Soaring Electronic discards Affects the Health of Millions of Children, WHO Warns.” June 15, 2021. https://www.who.int/news/item/15-06-2021-soaring-electronicdiscards-affects-the-health-of-millions-of-children-who-warns.

Wilkinson, J. C. 1996. “The Zanzibar Delimitation Commission 1885–1886.” *Geopolitics and International Boundaries* 1 (2): 130–58. https://doi.org/10.1080/13629379608407563.

Woodburn, James. 1982. “Egalitarian Societies.” *Man* 17 (3): 431–51. https://doi.org/10.2307/2801707.

Yanagisako, Sylvia Junko. 2003. *Producing Culture and Capital: Family Firms in Italy*. New Jersey: Princeton University Press.

Yeager, Rodger. 2020. *Tanzania: An African Experiment*. 2nd ed. New York: Routledge. https://doi.org/10.4324/9780429308215.

Zalasiewicz, Jan, Mark Williams, Colin N. Waters, Anthony D. Barnosky, John Palmesino, Ann-Sofi Rönnskog, Matt Edgeworth, et al. 2017. "Scale and Diversity of the Physical Technosphere: A Geological Perspective." *The Anthropocene Review* 4 (1): 9–22. https://doi.org/10.1177/2053019616677743.

Zeng, X., H. Duan, F. Wang, and J. Li. 2017. "Examining Environmental Management of Electronic Discards: China's Experience and Lessons." *Renewable Sustainable Energy Review* 72: 1076.

Zeng, Xianlai, John A. Mathews, and Jinhui Li. 2018. "Urban Mining of E-Waste Is Becoming More Cost-Effective Than Virgin Mining." *Environmental Science & Technology* 52 (8): 4835–41. https://doi.org/10.1021/acs.est.7b04909.

Index

About the Author

Hailing from Tanzania, ethnographer Samwel Moses Ntapanta studies contemporary urbanization in the western Indian Ocean coast. Colonialism, consumerism, the debris of late capitalism, economies of repairing, reusing and recycling, and the anthropology of money are some of the topics that he investigates in his work. In East Africa, he has spent the better part of a decade studying how science, technology, and society interact with one another.

www.ingramcontent.com/pod-product-compliance
Lightning Source LLC
LaVergne TN
LVHW011815200926
848484LV00005B/1203

* 9 7 8 1 6 6 6 9 5 6 2 6 9 *